中国地质大学(武汉)中央高校教改基金(本科教学工程)项目(2019G51)资助

KVM 及 Docker 虚拟化平台实验教程

KVM JI DOCKER XUNIHUA PINGTAI SHIYAN JIAOCHENG

吴湘宁　彭建怡　罗勋鹤　常　虹
黄燕霞　刘文中　陈　苗　邓玉娇　编著
代　刚　邓中港　李佳琪　王　稳

图书在版编目(CIP)数据

KVM 及 Docker 虚拟化平台实验教程/吴湘宁等编著.—武汉:中国地质大学出版社,2021.4
ISBN 978-7-5625-4966-6

Ⅰ.①K…
Ⅱ.①吴…
Ⅲ.①虚拟处理机-高等职业教育-教材 ②Linux 操作系统-程序设计-高等职业教育-教材 ③计算机网络-高等职业教育-教材
Ⅳ.①TP338 ②TP316.85 ③TP393

中国版本图书馆 CIP 数据核字(2020)第 270161 号

KVM 及 Docker 虚拟化平台实验教程		吴湘宁 等编著
责任编辑:王 敏	选题策划:王 敏	责任校对:徐蕾蕾
出版发行:中国地质大学出版社(武汉市洪山区鲁磨路 388 号)		邮政编码:430074
电 话:(027)67883511	传 真:(027)67883580	E-mail:cbb@cug.edu.cn
经 销:全国新华书店		http://cugp.cug.edu.cn
开本:787 毫米×1 092 毫米 1/16		字数:185 千字 印张:8
版次:2021 年 4 月第 1 版		印次:2021 年 4 月第 1 次印刷
印刷:武汉市珞南印务有限公司		印数:1—500 册
ISBN 978-7-5625-4966-6		定价:24.00 元

如有印装质量问题请与印刷厂联系调换

前 言

虚拟化技术可将一台计算机虚拟为多台逻辑计算机,每个逻辑计算机可运行不同的操作系统,不但应用程序都可以在相互独立的空间内运行而互不影响,同时还可显著提高计算机的工作效率。

由于虚拟化技术使用软件的方法重新定义划分计算机资源,可以实现计算机资源的动态分配、灵活调度、跨域共享,可提高计算机资源的利用率,降低成本,加快部署,极大增强系统整体的安全性和可靠性,使计算机资源能够真正成为社会基础设施,服务于各行各业灵活多变的应用需求,因此,虚拟化技术成为了实现云计算平台的核心技术。

随着虚拟化和云计算技术在不同领域的应用越来越广,社会对掌握虚拟化和云计算技术人才的需求也日益增加。虚拟化和云计算产品的安装、配置、管理及应用的技术人员需要具有较强的动手实践能力,为此,许多高校开设了虚拟化及云计算的课程,目的就是希望能够通过理论讲解及实验环节,培养学生安装、配置、管理、使用虚拟化系统的实际动手能力,从而满足社会对虚拟化系统运维及开发人才的需求。

本书是一本专门针对KVM及Docker两种普遍使用的开源虚拟化平台的实验指导书,目的是通过一些实验过程,让学生能够由浅入深地逐步掌握KVM及Docker的安装、配置、管理及应用开发过程。全书大致分为3个部分:第一部分介绍一些虚拟化实验必备的基础知识,包括虚拟化技术的基本定义、目的、分类和发展历史,以及云计算的特点、体系结构及服务模式,同时还说明了虚拟化和云计算技术之间的关系;第二部分着重介绍KVM的安装、配置,以及KVM环境下虚拟机开关机、虚拟机克隆、虚拟机快照、虚拟机动态迁移等管理方法;第三部分则介绍近年来比较流行的开源轻量化虚拟平台Docker的安装及使用方法,包括Docker环境的安装、配置、容器启动停止、容器导入导出、容器镜像管理,Docker容器中安装Linux系统和MySQL、Python等应用的方法,以及3种常见的Docker集群管理工具。

本书的主要内容如下。

第1章:虚拟化与云计算,介绍虚拟化与云计算的基本概念,以及虚拟化与云计算的关系。

第2章:KVM安装与配置,包括KVM环境安装、使用virt-manager连接和管理KVM虚拟机、在KVM上创建虚拟机的步骤和方法。

第3章:KVM虚拟机的管理,介绍使用virt-manager对KVM虚拟机进行管理的方法,包括虚拟机的开机、关机、重启、虚拟机克隆、虚拟机快照、虚拟机动态迁移的步骤和方法。

第 4 章:Docker 应用容器引擎开发,介绍 Docker 容器的配置、管理及应用开发方法。在概述 Docker 应用容器引擎概念的基础上,详细说明了 Docker 安装配置、Docker 容器镜像基础操作、Docker 容器安装操作系统、Docker 容器安装 MySQL 数据库、Docker 安装 Python 的步骤和方法,最后介绍了 Docker Machine、Docker Compose 和 Swarm 三种 Docker 集群工具的使用方法。

本书适合作为高校开展虚拟化及云计算相关课程的实验指导用书,也可作为初学 KVM 及 Docker 技术人员的参考用书。

本书由吴湘宁、彭建怡、罗勋鹤、常虹、黄燕霞、刘文中编写,硕士研究生陈苗、邓玉娇、代刚、邓中港、李佳琪、王稳参加了教材的校验和整理。教材在编写过程中,得到了中智讯(武汉)科技有限公司、北京红亚华宇科技有限公司的大力支持,在此表示感谢!

由于时间仓促,加之水平有限,书中难免会有错误和不妥之处,敬请读者批评指正。

吴湘宁
2021 年 1 月于湖北武汉

目 录

1 虚拟化与云计算 (1)
    1.1 虚拟化概述 (1)
    1.2 云计算概述 (5)

2 KVM 安装与配置 (10)
    2.1 安装 KVM 环境 (10)
    2.2 利用 virt-manager 连接 KVM 宿主机 (17)
    2.3 在 KVM 上建立第一台虚拟机 (21)

3 KVM 虚拟机的管理 (32)
    3.1 虚拟机基本操作与克隆 (32)
    3.2 虚拟机快照 (38)
    3.3 虚拟机的动态迁移 (42)

4 Docker 应用容器引擎开发 (58)
    4.1 Docker 应用容器引擎认识及环境安装实验 (58)
    4.2 Docker 基础操作实验 (67)
    4.3 Docker 应用开发实验 (89)
    4.4 Docker 三剑客认识及环境安装实验 (97)

主要参考文献 (117)

附录 CentOS 中 NFS(网络文件系统)的配置方法 (118)

1 虚拟化与云计算

在开展虚拟化实验之前,必须先了解关于虚拟化与云计算的一些重要概念。本章将介绍虚拟化、云计算以及虚拟化与云计算的关系。

1.1 虚拟化概述

1.1.1 虚拟化的定义

虚拟化的本质是将现有的计算机资源通过虚拟的技术分割成若干个逻辑计算机资源,这些逻辑计算机资源相互独立。虚拟化的最终目标是提高计算机的利用效率,使计算机资源使用的灵活性最大化。虚拟化代表着IT应用未来的发展趋势,其模型如图1-1所示。

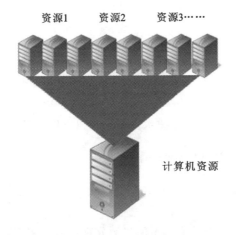

图1-1 单个计算机资源分割成多个逻辑资源

1.1.2 虚拟化的目的

虚拟化的主要目的是对IT基础设施和资源管理方式进行简化,帮助企业减少IT资源的开销,整合资源,节约成本。从近些年虚拟机被大量部署到企业的成功案例可以看出,越来越多的企业开始关注虚拟化技术给企业带来的好处,同时也在不断地审视自己目前的IT基础架构,从而希望改变传统架构。根据虚拟化技术的特点,其应用价值可以体现在"云"办公、虚拟制造、工业、金融业、政府和教育机构等方面。

虚拟化解决了许多当今遇到的问题，主要体现在以下5个方面。

（1）在一个特定的软硬件环境中虚拟另一个不同的软硬件环境，从而打破层级依赖。

（2）提高计算机设备的利用率。在一台物理服务器上同时安装并运行多种操作系统，从而提高物理设备的使用率。

（3）增强整体的安全性和可靠性，实现故障隔离，当一台虚拟机发生故障时，不会影响其他虚拟机及宿主机的操作系统。

（4）虚拟化可以统一虚拟资源而达到融合的目的，使不同品牌的硬件相互兼容。虚拟资源独立于硬件，无需修改即可在不同的物理服务器之间迁移。

（5）在硬件采购、电力消耗、机房温度控制和服务器机房空间等方面都可节约潜在成本，如表1-1所示。

表1-1 虚拟化节约潜在成本

类别	可节约的潜在成本
硬件	不需要为每台服务器桌面配置硬件
电力消耗	每台物理机消耗的电力是一定的，不会随着虚拟机规模的增长而增长
机房温度控制	无需添加新的制冷设备
服务器机房空间	虚拟机不是物理机器，所以无需增加数据中心空间

1.1.3 虚拟化的分类

1.1.3.1 按使用目的分类

从虚拟化的使用目的来看，虚拟化技术主要分为以下四大类。

1. 平台虚拟化（Platform Virtualization）

平台虚拟化是指针对服务器和操作系统的虚拟化，主要包括服务器虚拟化和桌面虚拟化。

服务器虚拟化是将一个操作系统的物理实例分割到虚拟实例或者虚拟机中，这些虚拟操作系统可以是×86或者×64的Windows、Linux或者UNIX操作系统。服务器虚拟化又分为软件虚拟化和硬件虚拟化。软件虚拟化是指在一个虚拟化平台上运行虚拟化操作系统，而这个虚拟化平台运行在现有的操作系统上，属于寄居架构，如图1-2所示，比如大家熟知的VMware Workstation。硬件虚拟化则是指虚拟化平台直接运行在物理硬件上，这种虚拟化通常又称为Hypervisor。Hypervisor运行在硬件系统之上、虚拟化操作系统之下，可实现对硬件资源的分割分配，属于原生架构。原生架构不需要操作系统，由Hypervisor直接管理硬件，如图1-3所示。

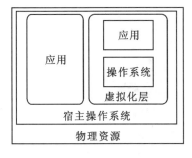
图 1-2 寄居架构

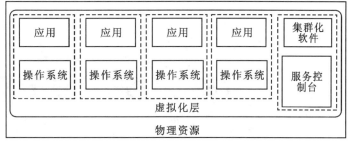
图 1-3 原生架构

桌面虚拟化是指将计算机的桌面进行虚拟化，这种虚拟化允许依靠虚拟机来提供系统桌面，以实现桌面使用的安全性和灵活性。用户可以通过任意设备，在任意地点、任意时间访问网络上属于自己的桌面系统。

2. 资源虚拟化（Resource Virtualization）

资源虚拟化主要是指虚拟计算机中的使用资源，包括存储虚拟化和网络虚拟化。存储虚拟化最通俗地讲就是对存储硬件资源进行抽象化表现，用于合并多个设备中的物理存储，使其表现为一个单一的"存储池"。对于用户来说，虚拟化的存储资源就像一个巨大的存储池，用户不会看到具体的磁盘、磁带，也不必关心自己的数据是经过哪一条路径通往哪一个具体的存储设备，例如 VMware 存储虚拟化架构，如图 1-4 所示。

网络虚拟化是将一条网络带宽分割成若干个相互独立的通道，以此来控制可用带宽，将可用带宽分配给特定的资源。比较常见的就是虚拟局域网，即在物理局域网内创建逻辑网络，而这两种网络互不影响。

3. 应用程序虚拟化（Application Virtualization）

基于软件的服务虚拟化是将应用程序从操作系统中分离出来，使应用程序运行在操作系统中，但是又不依赖于操作系统。应用程序虚拟化为应用程序提供了一个虚拟的运行环境，在这个环境中，不仅包括应用程序的可执行文件，还包括它所需要的运行时的环境。

4. 表示层虚拟化（Present Virtualization）

用户在使用应用程序时，其应用程序并不是运行在本地操作系统之上的，而是运行在服务器上面的，客户机只显示程序的界面和用户的操作，服务器仅向用户提供表示层，这种虚拟化就是表示层虚拟化。

1.1.3.2 按虚拟化技术分类

若从虚拟化的技术来划分，虚拟化又分为全虚拟化和半虚拟化。

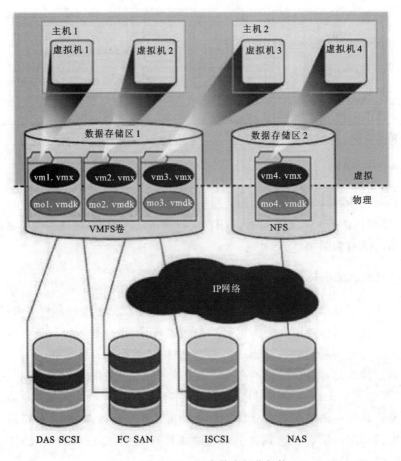

图 1-4　VMware 存储虚拟化架构

1. 全虚拟化

全虚拟化为客户机提供了完整的虚拟×86平台，包括处理器、内存等其他资源，支持运行绝大多数的操作系统，为虚拟机的配置提供了最大程度的灵活性。不需要对客户机操作系统进行任何修改，即可正常运行任何非虚拟化环境中基于×86平台的操作系统和软件。VirtualBox、KVM、VMware 均采用全虚拟化模式，如图 1-5 所示。

2. 半虚拟化

半虚拟化也叫准虚拟化，它也使用 Hypervisor 分享存取底层的硬件，但是它的客户机操作系统中也集成了虚拟化方面的代码。客户机操作系统意识到自己是处于虚拟化环境，因为操作系统自身能够与虚拟进程进行很好的协作。半虚拟化需要客户操作系统做一些修改（以配合 Hypervisor）。半虚拟化提供了与原始系统相近的性能。与全虚拟化一样，半虚拟化可以同时支持多个不同的操作系统。Xen 采用了半虚拟化模式，如图 1-6 所示。

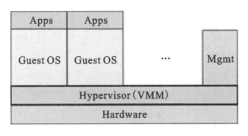
图1-5 全虚拟化架构

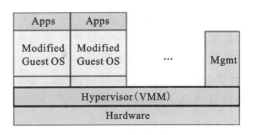
图1-6 半虚拟化架构

1.1.4 虚拟化的历史

在当今社会,世界各地的数据中心都在研究虚拟化技术,希望以此来提高数据中心的工作效率。虚拟化并不是今天才有的技术,而是经过了漫长的发展历程。现在,虚拟化不但是一门很热门的技术,还是一门各大企业在追求企业效率和信息化时期望最高的技术。

虚拟化的发展大体可以分为以下4个阶段。

提出概念:虚拟化概念是在1959年6月,由计算机科学家Christopher Strachey在国际信息处理大会(International Conference on Information Processing)上发表的论文《大型高速计算机中的时间共享》(*Time Sharing in Large Fast Computer*)中首次提出。

开发技术:从20世纪60年代开始,IBM的操作系统虚拟化技术使得大型机的资源得到充分利用。同时,IBM还推出了支持虚拟化的小型机,如IBM 360/40、IBM 360/67、VM/370等。

蓬勃发展:到了20世纪90年代,由VMware公司率先实现了×86服务器架构上的虚拟化,并在1999年推出了×86平台上的第一款虚拟化商业软件VMware Workstation,从而加快了虚拟化的发展脚步。

群雄逐鹿:随着虚拟化在×86平台上的发展,其带来的低成本等诸多好处,促使更多的厂家加入了虚拟化技术的开发队伍,同时也出现了很多支持虚拟化的产品,如Windows操作系统下的Virtual PC、Parallels的Workstation以及VirtualBox等,这些虚拟化产品最后都被大厂商收购。国外市场占有率较高的厂家有VMware、Citrix(思杰)、Microsoft。开源的虚拟化平台,如KVM、Docker、OpenStack等,使得虚拟化技术得到了进一步推广应用。尤其值得一提的是Docker,是基于Linux Container(容器)的轻量级虚拟化技术,其特点是启动快、资源占用小,它适合用来构建自动化测试和持续集成环境,以及一些需要横向扩展的应用(如需要快速启停来应对峰谷的Web应用)。近年来,一些国内厂商基于开源技术推出的虚拟化云平台占据了不少市场份额,如华为的FusionSphere云操作系统、新华三的H3C CAS虚拟化平台等。

1.2 云计算概述

云计算是通过互联网将某一计算任务分布到大量的计算机上,并可配置共享计算的资源池,且共享软件资源和信息资源可以按需提供给用户和设备的一种技术。通过云计算技术,我们只需要一台笔记本电脑或者一部手机,就可以通过网络来获取一切数字资源,甚至包括超级计算这样的任务。

其实云计算这个名词的提出比虚拟化概念的提出要晚很多。2006年8月9日,Google首席执行官埃里克·施密特(Eric Schmidt)在搜索引擎大会(SES San Jose 2006)上首次提出"云计算"(Cloud Computing)的概念。Google 云计算源于 Google 工程师克里斯托弗·比希利亚所做的 Google 101 项目。

1.2.1 云计算的特点

云计算主要有以下5个特点。

1. 基于互联网络

云计算可以把一台一台的服务器用网络连接起来,使它们相互之间可以进行数据传输。数据通过网络像云一样自动"飘到"另一台服务器上。云计算同时通过网络向用户提供服务。

2. 按需服务

"云"的规模可以动态伸缩。用户在使用云计算服务的时候,是按照自己所能承受的费用获得计算机服务资源的。这些计算机服务资源会根据用户的个性化需求增减,或者通过云计算得到更多层次的服务,以满足不同用户的需求。

3. 资源池化

资源池是一种配置机制,是将所使用的各种资源(如网络资源、存储资源等)统一进行配置,用户无需关心这些资源采取的设备型号、复杂的内部结构、实现的方法和地理位置。从用户的角度来看,这些资源是一个整体的设备,可按需为用户提供服务。作为这些资源的管理者来说,资源池可以无限地增减和更换设备,统一管理、调度这些资源,使用户得到满足。

4. 高可用

云计算必须要保证服务的可持续性、安全性、高效性和灵活性,故其必须采用各种冗余机制、备份机制、足够完全的安全管理机制、高效的反应机制和保证存取海量数据的灵活机制等,从而保证用户数据和服务的安全可靠。

5. 资源可控

云计算提出的初衷,是让人们能够像使用水电一样便捷地使用云计算服务,极大地方便人们获取计算服务资源,并有效节约技术成本,使计算资源的服务效益最大化。事实上,在云计算在线计费服务领域,如何对云计算服务进行合理和有效的计费,即如何就提供的云计算服务向最终用户收取服务费用,仍然是一项值得业界关注的课题。

1.2.2 云计算体系结构

从技术的角度来看,业界通常认为云计算体系分为3个层次,包括基础设施即服务

(IaaS,Infrastructure as a Service)、平台即服务(PaaS,Platfrom as a Service)和软件即服务(SaaS,Software as a Service)。这3层服务对于用户来说是相互独立的,因为每层提供的服务各不相同。但从技术角度来看,3层服务是相互依赖的,但是不相互依存。云计算的体系结构如图1-7所示。

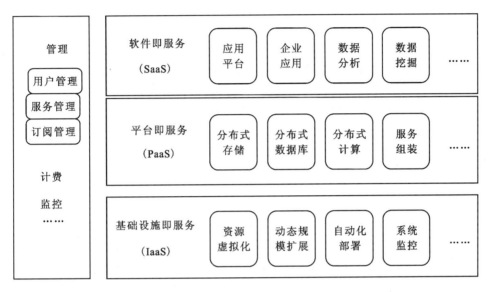

图1-7 云计算的体系结构

1. IaaS(基础设施即服务)

这一层的作用是将各个底层的计算和存储等资源作为服务提供给用户。用户能够部署和运行任意软件,包括操作系统和应用程序。用户不能管理或控制任何云计算基础设施,但能控制操作系统的选择、空间的存储、部署的应用,也有可能获得有限制的网络组件的控制。

2. PaaS(平台即服务)

简单地说,PaaS平台就是指云环境中的应用基础设施服务,也可以说是中间件即服务。PaaS是服务提供商提供给用户的一个平台,用户可以在这个平台上利用各种编程语言和工具(如Java、Python、.NET等)开发自己的软件或者产品,并且部署应用和应用的环境,而不用关心其底层的设施、网络、操作系统等。

3. SaaS(软件即服务)

SaaS提供商为用户搭建了信息化所需要的所有网络基础设施及软件、硬件运作平台,并负责所有前期的实施、后期的维护等一系列服务。而用户只需要通过终端,以Web访问的形式来使用、访问、配置各种服务,而不用管理任何在云计算上的服务。

1.2.3 云计算的模式

云计算的模式种类有很多种,按照云计算的服务模式主要分为 4 种,分别是公有云、私有云、混合云和行业云。

1. 公有云

公有云通常指第三方提供商为用户提供的能够使用的云,或者是企业通过自己的基础设施直接向外部用户提供服务的云。在这种模式下,外部用户可以通过互联网访问服务,但不拥有云计算资源。用户使用公有云可能是免费的或成本相对低廉的。这种云可在当今整个开放的公有网络中提供服务。世界主要的公有云有微软的 Windows Azrue、Google 的 Apps、Amazon 的 AWS。国内的公有云有阿里云、腾讯云、华为云等。公有云具有费用较低、灵活性高、可大规模应用等优点。

2. 私有云

私有云通常是指用户自己开发或者使用云计算产品自己搭建(也可由云提供商构建)云计算环境并只为自己提供服务的云计算。私有云是为单独使用而构建的,因而可提供对数据、安全性和服务质量的最有效控制。私有云具有数据安全性高、能充分利用资源、服务质量高等优点。

3. 混合云

对于信息控制、可扩展性、突发需求以及故障转移需求来说,只有将公有云和私有云相结合才可满足,这种两者结合起来的云就是混合云。其优势是用户可以享受接近私有云的私密性和接近公有云的成本,并且能快速地接入大量位于公有云的计算能力,以备不时之需。

4. 行业云

顾名思义,行业云是针对某个行业设计的云,并且仅开放给这个行业内的企业。行业云是由我国著名的商用 IT 解决方案提供商浪潮提出的。行业云由行业内或某个区域内起主导作用或者掌握关键资源的组织建立和维护,并以公开或者半公开的方式,向行业内部或相关组织和公众提供有偿或无偿服务。

1.2.4 云计算与虚拟化的关系

云计算与虚拟化之间有什么关系?虚拟化为云计算提供了很好的底层技术平台,而云计算是最终产品。虚拟化重点在于对资源的虚拟,比如把一台大型的服务器虚拟成多台小型的服务器,是一种侧重虚拟的技术。而云计算的重点是对资源池(可以是经过虚拟化之后)进行统一的管理和调度,是一种侧重于对虚拟化之后的资源进行管理和调度的技术。

虚拟化是云计算的主要支撑技术之一。虚拟化将应用程序和数据在不同层次以不同的面貌展现,这样有助于使用者、开发人员和维护人员方便地使用、开发与维护这些应用程序及数据。虚拟化允许 IT 部门添加、减少移动硬件和软件到它们想要的地方。虚拟化为组织带来灵活性，从而改善 IT 运维和减少成本支出。但是虚拟业务的运行、资源的分类仍然需要以云计算作为总体管理框架，通过云平台高效简捷的管理流程，降低虚拟资源池的运维和管理成本，从而最大化地实现虚拟化的巨大价值。

云计算和虚拟化是密切相关的，但是虚拟化对于云计算来说并不是必不可少的，也就是说云计算管理的资源并不仅限于虚拟资源，也可能还包含一些实体资源。

云计算将各种 IT 资源以服务的方式通过互联网交付给用户。然而虚拟化本身并不能给用户提供自服务层。没有自服务层，就不能提供计算服务。云计算模型允许终端用户自行申请自己的服务器、应用程序和包括虚拟化资源在内的各类资源，这使得企业可以最大限度地利用自身的计算资源。

虽然虚拟化和云计算并非捆绑技术，但二者可以通过优势互补为用户提供更优质的服务。云计算方案使用虚拟化技术使整个 IT 基础设施的资源部署更灵活。反过来，虚拟化方案也可以引入云计算的理念，为用户提供按需使用的资源和服务。在一些特定业务场景中，云计算和虚拟化是很难区分的，只有同时应用这两项技术，才能够顺利开展服务。

2 KVM 安装与配置

2.1 安装 KVM 环境

2.1.1 实验目的

(1)认识 KVM。
(2)掌握 KVM 环境的安装与配置。

2.1.2 实验环境

(1)硬件环境:计算节点(无硬盘保护卡)1 台。
(2)软件环境:CentOS 系统、KVM 虚拟化软件。

2.1.3 实验内容

根据 KVM 虚拟化的特性,本节实验主要包含以下 3 个内容:①准备安装环境;②安装宿主机 Linux 系统;③KVM 的安装与配置。

2.1.4 实验步骤

KVM 是一个开源的系统虚拟化模块,自 Linux 2.6.20 之后集成在 Linux 的各个主要发行版本中。它使用 Linux 自身的调度器进行管理,所以相较 Xen,其核心源码很少。KVM 的虚拟化需要硬件支持(如 Intel VT 技术或者 AMD V 技术),是基于硬件的完全虚拟化。KVM 的核心部分为 KVM 模块和 QEMU,KVM 模块实现了硬件虚拟化的功能,QEMU 则实现了虚拟机的创建等功能。QEMU 提供了丰富的虚拟机管理的命令来配置、创建、启动虚拟机,这些命令具有各种各样的配置参数,若使用 QEMU 命令来管理虚拟机,对于一个新手而言比较困难,因此更简单方便的管理工具变得尤为重要。本节将介绍 KVM 的管理工具,这些管理工具都是基于 KVM 模块和 QEMU 两大核心功能,给虚拟机的操作以及虚拟机的管理带来极大的便利。KVM 管理工具有很多,如 virt-manager、libvirt、virsh、virt-viewer、virt-install、virt-top 等软件,本章重点介绍 virt-manager。

一、准备安装环境

KVM主机是以虚拟机形式提供虚拟机服务，安装KVM对硬件资源有一定的要求，硬件资源要求如下：①必须支持HVM（启用Intel VT或AMD V）；②64位×86架构CPU（多核CPU性能更佳）；③硬件虚拟化支持；④4GB内存；⑤36GB本地磁盘空间；⑥至少1块网卡。

准备好了硬件资源环境后，就需要准备KVM安装所需的系统镜像以及相应的软件资源。KVM是基于内核的虚拟化技术，要运行KVM虚拟化环境，就需要安装1个Linux操作系统。Linux系统目前有很多版本，安装KVM所需的Linux系统可以选择RHEL、CentOS、Fedora、Ubuntu等系统，在本书中KVM宿主机的操作系统选择CentOS 7。准备好宿主机系统之后，还要准备KVM安装相应的软件包，如KVM、QEMU、libvirt等，而这些软件的安装包一般在发行版的系统光盘中可以找到。

在安装KVM环境之前需确保计算节点已经开启Inter VT功能，查看方法：进入计算节点的BIOS设置界面，进入CPU设置选项，将虚拟化功能参数（一般显示为Virtualization Technology）设置成enabled即开启，然后保存BIOS设置。

二、安装宿主机Linux系统

在本实验中选用CentOS 7作为KVM宿主机的操作系统，并以root用户登录。

三、KVM的安装与配置

安装完KVM宿主机的操作系统之后，就要开始安装KVM相关的应用程序，如KVM、QEMU和libvirt等。

1. 检查宿主机的CPU是否支持虚拟化

```
[root@centos7 ~]# grep vmx /proc/cpuinfo
```

如果有"vmx"的信息输出，就说明支持宿主机VT。如果没有任何的输出，说明宿主机的CPU不支持虚拟化，将无法使用KVM虚拟机，此时，要确保BIOS里面已经开启虚拟化功能。

2. 查看是否已加载KVM模块

```
[root@centos7 ~]# lsmod | grep kvm
kvm_intel              162153  0
kvm                    525259  1 kvm_intel
```

如果没有加载，则运行以下命令加载KVM模块：

```
[root@centos7~]# modprobe kvm
[root@centos7~]# modprobe kvm-intel
[root@centos7~]# lsmod | grep kvm
```

内核模块导出了一个名为"/dev/kvm"的设备,这个设备将虚拟机的地址空间独立于内核或者任何应用程序的地址空间,可用以下命令查看一下:

```
[root@centos7 ~]# ll /dev/kvm
crw-rw-rw-+ 1 root kvm 10, 232 12月 28 16:13 /dev/kvm
```

3. 桥接网络

试一下网桥管理工具 brctl,如果命令执行不了,则需要安装 bridge-utils,并重启网络:

```
[root@centos7~]# yum -y install bridge-utils
[root@centos7~]# systemctl restart network
```

配置 KVM 的网桥模式:

```
[root@centos7~]# cd /etc/sysconfig/network-scripts/
[root@centos7 network-scripts]# cp ifcfg-eno16777736 ifcfg-br0
```

编辑网桥的设置文件:

```
[root@centos7 network-scripts]# gedit ifcfg-br0
TYPE="Bridge"           #这一行中,类型要修改为 Bridge
BOOTPROTO="dhcp"
DEFROUTE="yes"
PEERDNS="yes"
PEERROUTES="yes"
IPV4_FAILURE_FATAL="no"
IPV6INIT="yes"
IPV6_AUTOCONF="yes"
IPV6_DEFROUTE="yes"
IPV6_PEERDNS="yes"
IPV6_PEERROUTES="yes"
IPV6_FAILURE_FATAL="no"
NAME="br0"              #修改设备名称为 br0
# UUID="f718c3b9-c756-4ebd-9edc-3586f9e79f6d"         #UUID这一行要注释掉
DEVICE="br0"            #修改设备为 br0
ONBOOT="yes"
```

编辑网卡的设置文件:

```
[root@centos7 network-scripts]# gedit ifcfg-eno16777736
TYPE="Ethernet"
BRIDGE=br0              #添加这一行,说明网卡所使用的网桥
BOOTPROTO="dhcp"
DEFROUTE="yes"
PEERDNS="yes"
```

```
PEERROUTES="yes"
IPV4_FAILURE_FATAL="no"
IPV6INIT="yes"
IPV6_AUTOCONF="yes"
IPV6_DEFROUTE="yes"
IPV6_PEERDNS="yes"
IPV6_PEERROUTES="yes"
IPV6_FAILURE_FATAL="no"
NAME="eno16777736"
UUID="f718c3b9-c756-4ebd-9edc-3586f9e79f6d"
DEVICE="eno16777736"          #这个是网卡的设备名
ONBOOT="yes"
```

重启网络,查看网络信息,并测试:

```
[root@centos7 network-scripts]# systemctl restart network
[root@centos7 network-scripts]# ifconfig
br0: flags=4163<UP,BROADCAST,RUNNING,MULTICAST> mtu 1500
     inet 192.168.1.13 netmask 255.255.255.0 broadcast 192.168.1.255
     inet6 fe80::20c:29ff:fe67:74b6 prefixlen 64 scopeid 0x20<link>
     inet6 240e:36d:cac:3800:20c:29ff:fe67:74b6 prefixlen 64 scopeid 0x0<global>
     ether 00:0c:29:67:74:b6 txqueuelen 0 (Ethernet)
     RX packets 33 bytes 2172 (2.1 KiB)
     RX errors 0 dropped 0 overruns 0 frame 0
     TX packets 31 bytes 5115 (4.9 KiB)
     TX errors 0 dropped 0 overruns 0 carrier 0 collisions 0
eno1: flags=4163<UP,BROADCAST,RUNNING,MULTICAST> mtu 1500
     ether 00:0c:29:67:74:b6 txqueuelen 1000 (Ethernet)
     RX packets 76781 bytes 42367411 (40.4 MiB)
     RX errors 0 dropped 35 overruns 0 frame 0
     TX packets 4961 bytes 437780 (427.5 KiB)
     TX errors 0 dropped 0 overruns 0 carrier 0 collisions 0
[root@centos7 network-scripts]# ping www.baidu.com
PING www.a.shifen.com (14.215.177.38) 56(84) bytes of data.
64 bytes from 14.215.177.38: icmp_seq=1 ttl=55 time=23.5 ms
```

4. 安装 libvirt 及 kvm

```
[root@centos7 ~]# yum -y install libcanberra-gtk2 qemu-kvm.x86_64 qemu-kvm-tools.x86_64 libvirt.x86_64 libvirt-cim.x86_64 libvirt-client.x86_64 libvirt-java.noarch libvirt-python.x86_64 dbus-devel virt-clone virt-manager libvirt libvirt-python
```

设置 libvirt 服务并启用 libvirt：

```
[root@kevin ~]# systemctl enable libvirtd
[root@kevin ~]# systemctl start libvirtd
```

5.修改主机名

(1)查询主机名。

```
[root@centos7 ~]# hostname --fqdn
hostname: Name or service not known
```

此时说明带域名的主机名还没有被设置。
(2)编辑文件"/etc/hosts"。在文件的末尾处添加如下内容。
＜本机 IP 地址＞ ＜主机名＞.＜域名＞别名
示例(两台 KVM 宿主机的 IP 及域名)：

```
[root@censos7 ~] gedit /etc/hosts
192.168.1.13 kvm1.zonesion.com kvm1
192.168.1.14 kvm2.zonesion.com kvm2
```

此处将两台 KVM 宿主机的名称分别设置成 kvm1 和 kvm2，域名设置成 zonesion.com，也可以分别用别名 kvm1 和 kvm2 访问这两台宿主机。

```
[root@censos7 ~] ping kvm1.zonesion.com
[root@censos7 ~] ping kvm1
[root@censos7 ~] ping kvm2.zonesion.com
[root@censos7 ~] ping kvm2
```

(3)编辑文件"/etc/hostname"及"/etc/sysconfig/network"，这里将宿主机 kvm1 的HOSTNAME 改成 kvm1。

```
[root@centos7 ~]# gedit /etc/hostname
kvm1
[root@censos7 ~] gedit /etc/sysconfig/network
NETWORKING=yes
HOSTNAME=kvm1
```

(4)修改"/etc/idmapd.conf"，将 Domain 前面的注释符"#"去掉，并设置域名。

```
[root@censos7 ~] gedit /etc/idmapd.conf
Domain = zonesion.com
```

(5)配置文件更改完成后，重启网络服务。

```
[root@censos7 ~]# service network restart
```

(6)通过命令"hostname --fqdn"重新检查主机名,此时应返回刚才修改之后的主机名,如果没有返回正确的主机名,重启之后会显示正确的主机名。

```
[root@kvm1 ~]# hostname --fqdn
kvm1.zonesion.com
```

6. SELINUX 安全设置

编辑文件"/etc/selinux/config",将 SELINUX 参数设置成"permissive":

```
[root@kvm1 ~]# gedit /etc/selinux/config
SELINUX=permissive
```

保存后需要系统重新启动方可生效,若要保持本次生效可执行下列命令:

```
[root@kvm1 ~]# setenforce permissive
```

7. NTP 时间同步设置

NTP 时间同步是为了保证相互之间通信的服务器保持一致的时间。

(1)安装 NTP。

```
[root@kvm1 ~]# yum install -y ntp
```

(2)配置 NTP 服务器。

默认情况下,CentOS 系统会将 NTP 服务器配置为如下内容:

```
server 0.centos.pool.ntp.org iburst
server 1.centos.pool.ntp.org iburst
server 2.centos.pool.ntp.org iburst
server 3.centos.pool.ntp.org iburst
```

用户可以根据自己的需求更改该 NTP 服务器的地址,用户实验环境若可以连接互联网,则可保持默认设置;若不能连接互联网,则按照下面步骤配置 NTP 服务器。

(3)修改 NTP 的配置文件。

```
[root@kvm1 ~]# gedit /etc/ntp.conf
```

在默认与 Internet 通信的 Server 端地址前添加注释符号"#":

```
#server 0.centos.pool.ntp.org iburst
#server 1.centos.pool.ntp.org iburst
#server 2.centos.pool.ntp.org iburst
#server 3.centos.pool.ntp.org iburst
```

添加 NTP 服务器的 IP 地址:

```
server 192.168.1.13
```

(4)将 NTP 服务设置成开机启动,并启动 NTP 服务。

```
[root@kvm1 ~]# chkconfig ntpd on
[root@kvm1 ~]# service ntpd start
```

如果重新配置,就需要输入"service ntpd restart"命令重启 ntpd 服务。

8. 配置 QEMU

KVM 只有一个相对简单的配置项,即修改 QEMU VNC 的参数配置。输入如下命令编辑文件"qemu.conf",找到"vnc_listen=0.0.0.0"那一行,将其前面的注释符"#"去掉,然后保存退出。

```
[root@kvm1 ~]# gedit /etc/libvirt/qemu.conf
vnc_listen=0.0.0.0
```

9. 配置 libvirt

KVM 使用 libvirt 来管理虚拟机。因此,正确地配置 libvirt 是非常重要的。为了实现虚拟机的动态迁移功能,libvirt 需要监听不可靠的 TCP 连接,另外还需要避免 libvirtd 尝试使用组播 DNS 进行广播。这些都是在"/etc/libvirt/libvirtd.conf"文件中进行配置。

(1)输入如下命令编辑 libvirtd.conf 配置文件。

```
[root@kvm1 ~]# gedit /etc/libvirt/libvirtd.conf
```

找到下列参数选项,将注释符号"#"去掉,并修改相应的参数值。

```
listen_tls = 0
listen_tcp = 1
tcp_port = "16509"
mdns_adv = 0
auth_tcp = "none"
```

(2)修改/etc/sysconfig/libvirtd 中的参数。

```
[root@kvm1 ~]# gedit /etc/sysconfig/libvirtd
```

将下面内容行的注释符"#"去掉。

```
#LIBVIRTD_ARGS="--listen"
```

(3)重启 libvirt。

```
[root@kvm1 ~]# service libvirtd restart
Redirecting to /bin/systemctl restart libvirtd.service
```

10. 防火墙配置

为了保证 KVM 与 KVM 宿主机、KVM 管理服务器之间的正常通信,KVM 宿主机的一些 TCP 端口就需要打开,打开的端口有:22(SSH);1798;16509(libvirt);5900-6100(VNC consoles);49152-49216(libvirt live migration)。

要打开上述 TCP 端口,只需执行下列命令即可。

```
[root@kvm1 ~]# iptables-I INPUT-p tcp-m tcp--dport 22-j ACCEPT
[root@kvm1 ~]# iptables-I INPUT-p tcp-m tcp--dport 1798-j ACCEPT
[root@kvm1 ~]# iptables-I INPUT-p tcp-m tcp--dport 16509-j ACCEPT
[root@kvm1 ~]# iptables-I INPUT-p tcp-m tcp--dport 5900:6100-j ACCEPT
[root@kvm1 ~]# iptables-I INPUT-p tcp-m tcp--dport 49152:49216-j ACCEPT
```

输完上述命令后,相应的 TCP 端口就打开了,为了让这些端口一直处于打开状态,则需要保存防火墙设置。

```
[root@kvm1 ~]# iptables-save > /etc/sysconfig/iptables
```

2.2 利用 virt-manager 连接 KVM 宿主机

2.2.1 实验目的

(1)掌握 virt-manager 的安装及使用。
(2)掌握利用 virt-manager 连接 KVM 宿主机的方法。

2.2.2 实验环境

(1)硬件环境:计算节点(已安装 KVM 相关环境)1 台,管理节点 1 台。
(2)软件环境:virt-manager。

2.2.3 实验原理

KVM 提供了一系列的管理虚拟机的命令,在 KVM 宿主机上通过输入这些命令就可以实现虚拟机的管理,由于输入的命令较多,而且界面不友好,所以基于图形化界面的虚拟机管理软件变得尤为重要。

virt-manager 意为虚拟机管理器(Virtual Machine Manager),其应用于管理虚拟机的图形化的桌面用户接口,目前仅支持在 Linux 系统或者其他 UNIX 系统中运行。virt-manager 是由 Redhat 公司发起的项目,在 RHEL 6.x、Fedora、CentOS 等 Linux 发行版中有着非常广泛的应用。virt-manager 使用了 Python 语言开发其应用程序部分,使用 GNU AutoTools(包括 autoconf、automake 等工具)进行项目的构建。virt-manager 是一个完全开源的软件,使用的是 Linux 界广泛采用的 GNU GPL 许可证发布。virt-manager 依赖的一些程序库主要包括 Python(用于应用程序逻辑部分的实现)、GTK+PyGTK(用于 UI 界面)和 libvirt(用于底层的 API)。

virt-manager 工具在图形界面中实现了一些丰富的虚拟化管理功能,具体功能如下。

(1)对虚拟机生命周期的管理,如创建虚拟机、编辑、启动、暂停、恢复和停止虚拟机,还包括虚拟快照、动态迁移等功能。

(2)运行中虚拟机的实时性能、资源利用率等的监控,统计结果的图形化展示。

(3)对创建虚拟机的图形化的引导,对虚拟机的资源分配和虚拟硬件的配置与调整等功能也提供了图形化的支持。

(4)内置了一个 VNC 客户端,可用于连接到虚拟机的图形界面进行交互。

(5)支持本地或者远程管理 KVM、Xen、QEMU、LXC 等虚拟化平台上的虚拟机。

2.2.4 实验内容

本节实验中主要内容为:①安装 virt-manager;②利用 virt-manager 连接 KVM 宿主机。

2.2.5 实验步骤

一、准备工作

(1)准备 1 台已经安装好图形化界面的 CentOS 7。

(2)确保带有图形化界面的 CentOS 系统可以联网或者已经搭建本地源(这样做的目的是确保在后面的步骤当中可以正常安装 virt-manager)。

(3)准备 1 台 KVM 宿主机。

二、安装 virt-manager

(1)登录到 CentOS 7 的图形用户界面中,在桌面上的左上角用鼠标选择"应用程序→工具→终端",进入到命令窗口界面。

(2)输入如下命令安装 virt-manager。

```
[root@kvm1 ~]# yum install -y virt-manager
```

(3)安装完成后,就可以在"应用程序→系统工具"中看到"虚拟系统管理器"了。

三、利用 virt-manager 连接 KVM 宿主机

virt-manager 安装结束后,若想对 KVM 宿主机进行管理,可以将 virt-manager 连接到 KVM 宿主机,连接方法如下。

(1)登录到 CentOS 7 的图形用户界面中,在桌面上的左上角用鼠标选择"应用程序→系统工具→虚拟系统管理器",或使用"virt-manager"命令打开虚拟系统管理界面(图 2-1)。

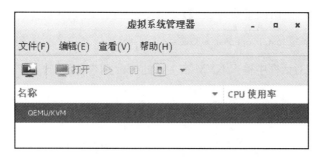

图 2-1 虚拟系统管理器

(2)在虚拟机管理软件上点击"文件→添加连接..."(图 2-2)。

图 2-2 添加连接

(3)在"添加连接"对话框中勾选"连接到远程主机",在"主机名"输入框中输入 KVM 宿主机的 IP 地址或主机名(图 2-3),勾选"自动连接",最后点击"连接"按钮。

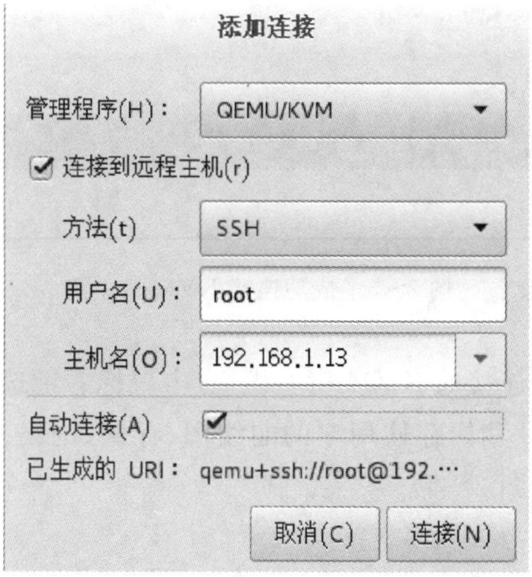

图 2-3 输入 KVM 宿主机的 IP 地址

如果是第一次连接主机,系统会弹出登录 SSH 的验证,输入"yes"即可。如果 KVM 主机还未安装 SSH,可以使用以下命令安装。

```
[root@kvm1 ~]# yum -y install openssh-askpass
```

(4)在"OpenSSH"对话框中输入 KVM 宿主机的 root 用户密码,然后点击"OK"按钮(图 2-4)。

图 2-4　输入 KVM 宿主机的密码

(5)连接成功后,系统会显示已经连接上这台 KVM 宿主机(图 2-5)。

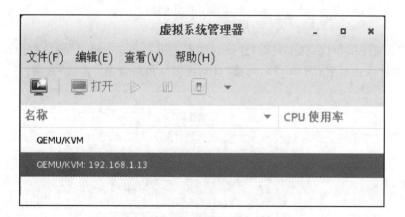

图 2-5　成功连接 KVM 宿主机

(6)在显示列表中选中连接的 KVM 宿主机,点击鼠标右键选择"详情"可以看到该 KVM 宿主机的硬件情况,如 CPU 和内存使用率的曲线图、存储、网络等信息,点击不同的选项卡,可以查看不同设备的信息(图 2-6)。

图 2-6　KVM 宿主机的硬件信息

2.3　在 KVM 上建立第一台虚拟机

2.3.1　实验目的

(1)熟悉 virt-manager 的使用。
(2)掌握在 KVM 上建立虚拟机的方法。

2.3.2　实验环境

(1)硬件环境:计算节点(已安装 KVM 相关环境)1 台,管理节点 1 台,存储节点 1 台。
(2)软件环境:virt-manager。

2.3.3　实验内容

在搭建好 KVM 相关环境之后,就可以利用 virt-manager 管理工具在 KVM 宿主机上建立虚拟机了,在 KVM 上可支持安装 Windows 系列、Linux 系列等大部分主流操作系统的虚拟机。

利用 virt-manager 在 KVM 上安装 VM(Virtual Machine,虚拟机)之前,需准备要安装的操作系统的 ISO 镜像,通常安装操作系统是利用本地的 ISO 镜像、光驱等方式进行安装,为了方便多组用户同时安装操作系统,在此建议用户先搭建一个 NFS 服务器,并在搭建好的 NFS 服务器下创建一个 ISO 存储池,将系统所需的 ISO 镜像全部上传到该 ISO 存储池中。当在 KVM 上安装 VM 时,只需要新建一个 ISO 存储池,并将该存储池指定到 NFS 服务器的 IP 地址以及 ISO 镜像的目录即可。

本节实验内容主要为如下几个内容:①创建 ISO 存储池;②创建第一台虚拟机。

2.3.4 实验步骤

一、创建 ISO 存储池

(1)搭建一个 NFS 服务器(在 CentOS 操作系统中设置 NFS 服务的方法详见附录,也可用其他操作系统实现 NFS 服务器),将 NFS 服务器上的防火墙关闭以保证能够顺利访问其开放的 NFS 目录,将系统安装所需的 ISO 镜像上传到 NFS 目录(如/export/ISO/)下。

(2)在 virt-manager 管理界面中选中 KVM 宿主机,点击鼠标右键选择"详情",然后进入"存储"选项卡页面,点击左下角的"+"(添加池)按钮添加存储池(图 2-7)。

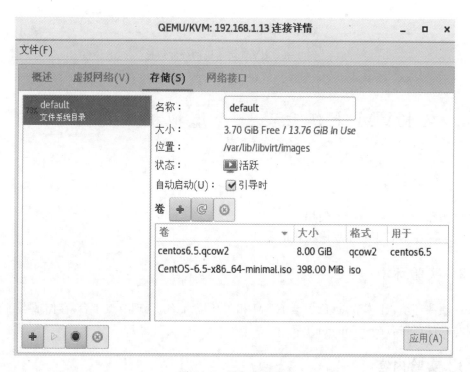

图 2-7　添加存储池

(3)输入存储池的名称,此处输入"NFS-ISO-library",选择存储类型为"netfs:网络导出的目录",然后点击"前进"进入下一步(图 2-8)。

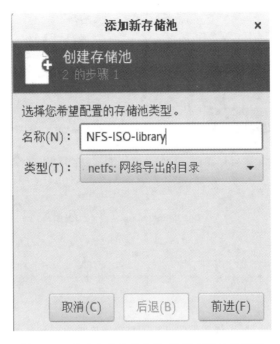

图 2-8 输入存储池的名称、选择存储池的类型

(4)在"主机名"中输入 NFS 服务器的 IP 地址,在"源路径"中输入 ISO 镜像文件在 NFS 中的存储路径,最后点击"完成"按钮即可完成 NFS 上 ISO 存储池的创建(图 2-9)。

图 2-9 配置存储池的信息

(5)ISO 存储池创建完成后,在"存储"选项左侧的列表中可看到新增加了一个 NFS ISO 存储池,点击该存储池,在右侧可以看到该存储池中所有 ISO 镜像(图 2-10)。

图 2-10　ISO 存储池成功创建

二、创建第一台虚拟机

ISO 存储池创建完成后就可以创建虚拟机了，下面介绍如何利用 virt-manager 在 KVM 宿主机上创建一个 CentOS 6.5 mini 版本操作系统的虚拟机。

(1)在 virt-manager 中选中列表中的 KVM 宿主机，点击鼠标右键选择"新建"(图 2-11)。

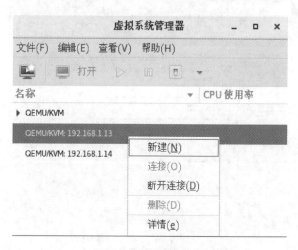

图 2-11　新建虚拟机

(2)选择虚拟机操作系统的安装方式。操作系统的安装方式有 4 种，分别是本地安装介质(ISO 映像或者光驱)、网络安装(HTTP、FTP 或者 NFS)、网络引导和导入现有磁盘映像。

本实验选择第一种安装方式,即本地安装介质,然后点击"前进"按钮进入下一步(图 2-12)。

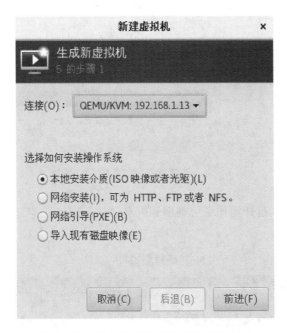

图 2-12 输入虚拟机名称、选择安装方式

(3)在新弹出的镜像选择对话框中勾选"使用 ISO 映像",然后点击"浏览…"按钮查找 ISO 文件的路径(图 2-13)。

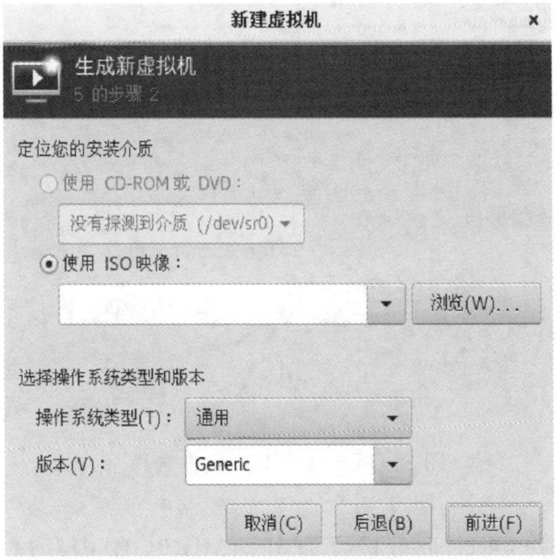

图 2-13 使用 ISO 镜像安装系统

在存储池列表中选中"NFS-ISO-library",在右侧出现的列表中选择所需的系统镜像,此处选择为"CentOS-6.5-mini.iso",最后点击"选择卷"(图 2-14)。

图 2-14 选择待安装的系统 ISO 镜像

ISO 镜像选择完毕后,选择操作系统类型和版本,然后点击"前进"进入下一步(图 2-15)。

图 2-15 完成 ISO 镜像选择

(4)配置虚拟机的内存、CPU,此处设置为 1024MB(1GB)内存,1 个 CPU,用户可根据自己的需求适当修改,然后点击"前进"按钮进入下一步(图 2-16)。

(5)配置虚拟机的磁盘大小为 8GiB。虚拟机磁盘配置有两个选项,系统会默认勾选第一个选项"为虚拟机创建磁盘镜像",在该模式下只要选择磁盘大小即可完成磁盘的创建,第二个选项是"选择或创建自定义存储",即可以将虚拟机放置在事先自定义好的存储卷上(图 2-17)。

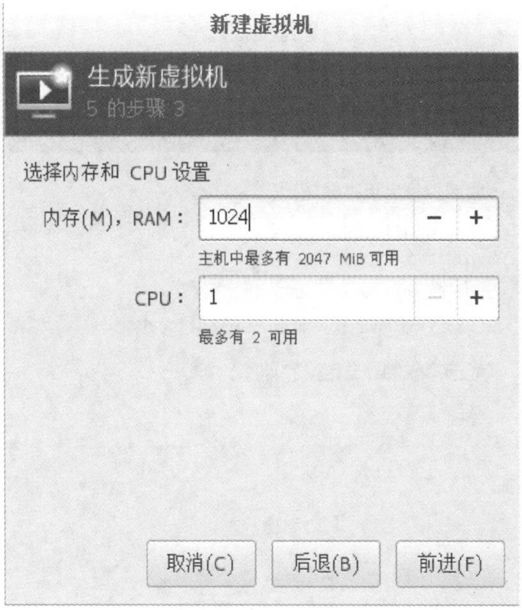

2 KVM 安装与配置

图 2-16 配置虚拟机的内存、CPU

图 2-17 第一种虚拟磁盘配置方式

下面是第二种磁盘方式配置过程。

第一，勾选"选择或创建自定义存储"，然后点击"管理..."按钮(图 2-18)。

第二，在弹出的页面中点击"＋"(创建新卷)按钮进入"创建存储卷"对话框。输入存储卷的名称，选择"格式"为"qcow2"，"最大容量"设置为 8GiB，然后点击"完成"按钮(图 2-19)。

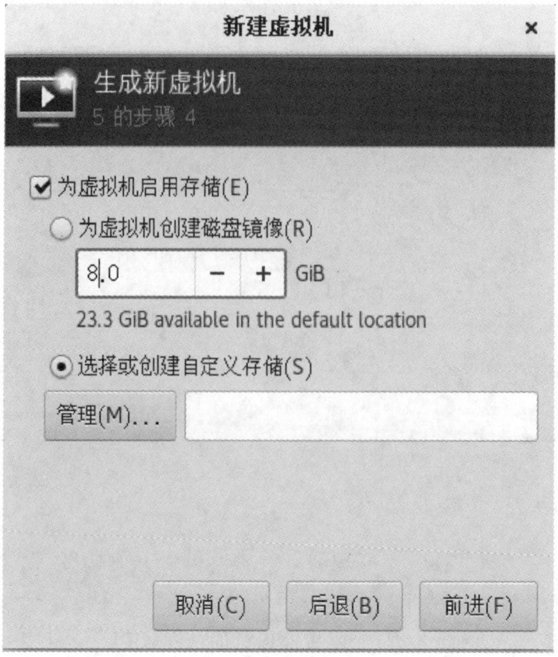

图 2-18　选择第二种磁盘配置方式

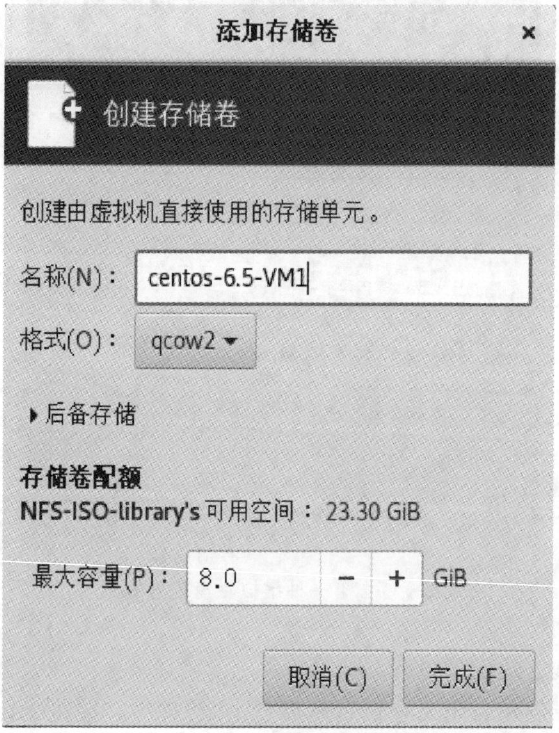

图 2-19　配置磁盘选项

第三,选中上一步骤创建的磁盘卷,点击"选择卷"按钮(图 2-20)。

图 2-20 选择创建好的磁盘卷

第四,点击"前进"进入下一步(图 2-21)。

图 2-21 完成磁盘配置

第五,完成虚拟机的硬件资源配置,点击"完成",开始虚拟机的创建(图 2-22)。
第六,完成虚拟机的创建后,虚拟机会启动进入系统安装的界面,如图 2-23 所示。

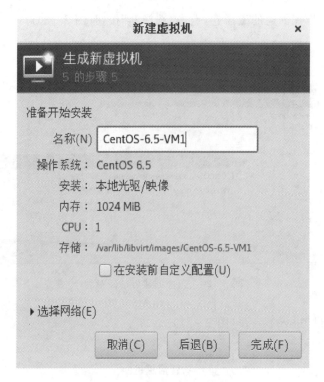

图 2-22 完成虚拟机的配置

图 2-23 虚拟机开始安装系统

第七，系统安装结束后，在 virt-manager 界面上可以看到 KVM 宿主机下面新增的虚拟机，如图 2-24 所示。

第八，在虚拟机的控制台界面，选中虚拟机，点击鼠标右键选择"打开"可打开虚拟机的控制台界面，如图 2-25 所示。

图 2-24 创建完成的虚拟机

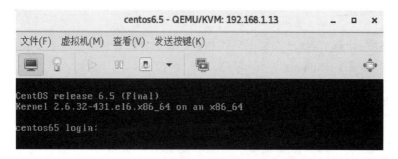

图 2-25 虚拟机的控制台界面

3 KVM 虚拟机的管理

3.1 虚拟机基本操作与克隆

3.1.1 实验目的

掌握 KVM 虚拟机的基本操作和克隆方法。

3.1.2 实验环境

(1)硬件环境:计算节点(已安装 KVM 相关环境)1台,管理节点1台,存储节点1台。
(2)软件环境:virt-manager。

3.1.3 实验内容

虚拟机安装好之后,就可以将其视作一台真实的物理机来操作。虚拟机的一些常用操作包括开机、关机、重启、控制台界面显示等。

克隆虚拟机对于多次部署虚拟机有很大帮助,由于安装部署一台虚拟机所耗费的时间比较长,所以虚拟机的克隆功能变得愈发重要。

3.1.4 实验步骤

一、虚拟机的常用操作

(1)虚拟机的开机、关机、重启操作。选中虚拟机,点击鼠标右键就可以选择虚拟机的重启、关机、强制关机等操作(图 3-1)。

或者是选中虚拟机后,直接点击"virt-manager"菜单栏中开机、关机按钮以及关机按钮旁边的下三角来选择虚拟机的重启、强制关机等操作(图 3-2)。

注意:用户在对 Linux 虚拟机执行开机、关机、重启等操作时,会发现虚拟机无反应。这是因为 virt-manager 在执行重启、关机时,向虚拟机下达了 virsh 命令的重启、关机指令,但 virsh 命令是通过调用 QEMU/KVM 的重启、关闭命令来重启、关闭虚拟机的,需要宿主机向虚拟机发送 ACPI 指令。然而 KVM 在安装 Linux 虚拟机时,默认情况下虚拟机中没有安装 acpid 程序,因此虚拟机无法接收并处理 ACPI 指令。要解决这个问题,只需在 Linux 虚拟机里面安装 acpid 服务,然后启动该服务即可。安装与配置过程如下。

3 KVM 虚拟机的管理

图 3-1 虚拟机的关机、重启等选项

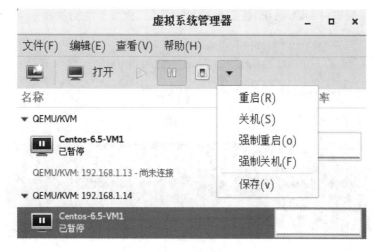

图 3-2 虚拟机的操作按钮

```
[root@CentOS-VM1 ~]# yum install -y acpid
[root@CentOS-VM1 ~]# service acpid start
[root@CentOS-VM1 ~]# chkconfig acpid on
```

(2)打开虚拟机控制台显示界面。选中虚拟机,点击鼠标右键选择"打开",或者点击菜单栏中的"打开"按钮(图3-3)。

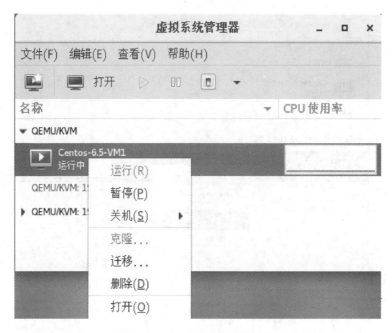

图3-3 打开虚拟机的控制台

图3-4是虚拟机(Linux系统)控制台的显示界面,相当于电脑的显示屏。

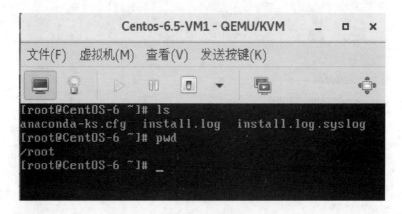

图3-4 虚拟机的控制台显示界面

(3)如果需要查看虚拟机的详细硬件配置信息,只需在虚拟机的控制台界面上点击"查看→详情",如图3-5所示。

进入虚拟机的详情页面后,就可以查看虚拟机的硬件配置信息,如图3-6所示。

3　KVM 虚拟机的管理

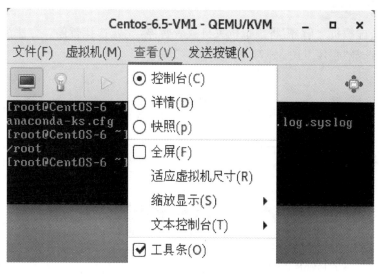

图 3-5　查看虚拟机的详细信息

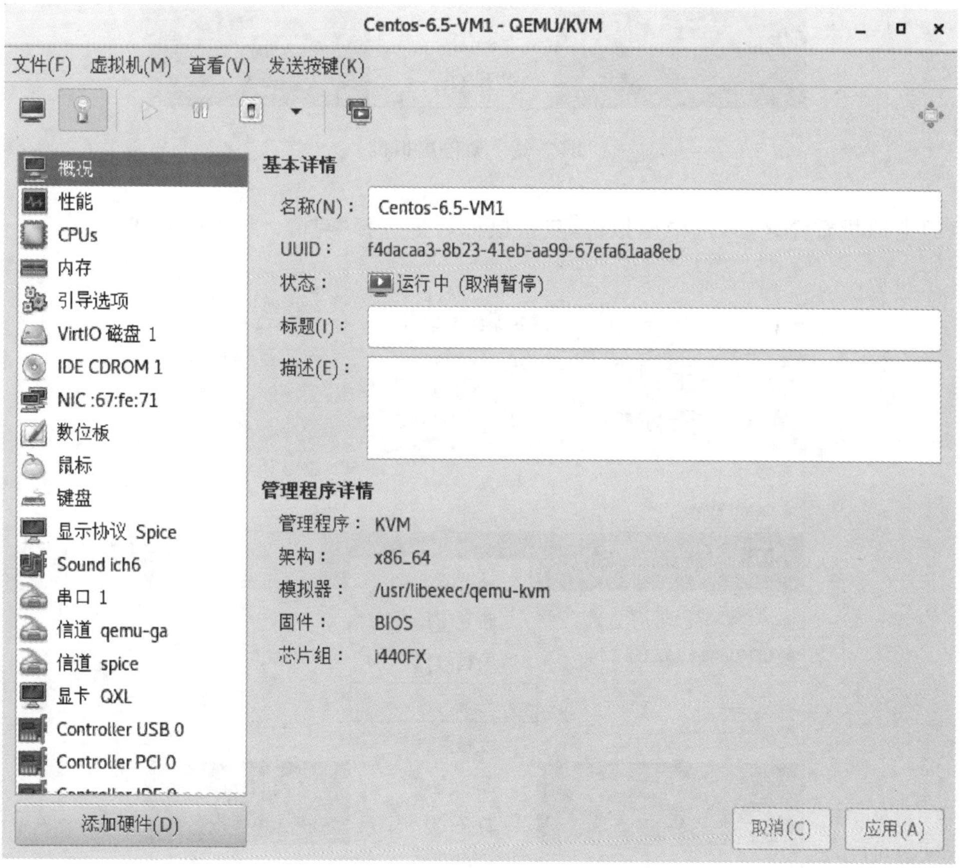

图 3-6　虚拟机的硬件配置信息

二、虚拟机的克隆

(1)在克隆虚拟机之前,需将虚拟机关机或者暂停,此处选择暂停。方法为:选中虚拟机,点击鼠标右键选择"暂停"(图3-7)。

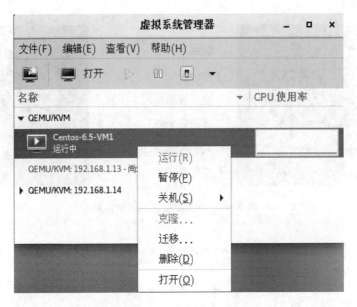

图3-7 暂停虚拟机

(2)虚拟机暂停之后,选中虚拟机,点击鼠标右键选择"克隆..."(图3-8)。

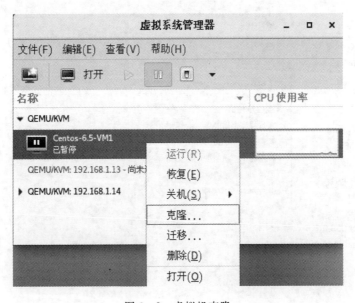

图3-8 虚拟机克隆

(3)编辑克隆虚拟机的名称,默认情况下,克隆的虚拟机名称会在原虚拟机名称后面加一个-clone 单词,点击"克隆"按钮开始克隆虚拟机(图 3-9)。

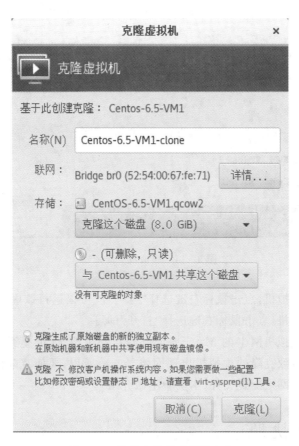

图 3-9 编辑克隆的虚拟机名称

(4)此时会出现克隆的进度条,克隆时间与虚拟机数据的大小有关,虚拟机数据越大,克隆时间则越长(图 3-10)。

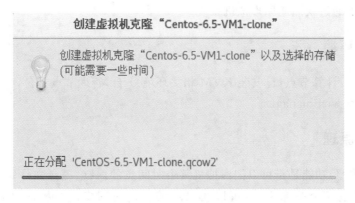

图 3-10 虚拟机克隆进度

(5)虚拟机克隆结束后,就会在虚拟机显示列表中显示克隆出的虚拟机(图3-11)。

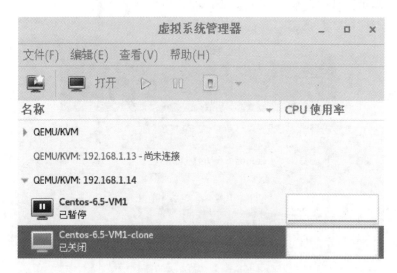

图3-11 虚拟机克隆成功

(6)选中克隆的虚拟机,点击鼠标右键选择"运行"就可以运行该虚拟机,要进入虚拟机的控制台,则选中该虚拟机,点击鼠标右键选择"打开"即可。

(7)克隆出来的虚拟机网卡的MAC地址已经自动更改了,但是系统的网卡配置文件却没有更改,这样就会导致克隆的虚拟机出现网络异常。所以,要使克隆的虚拟机网卡正常工作,用户还需修改相关网络配置文件(本实验不做介绍)。

3.2 虚拟机快照

3.2.1 实验目的

(1)掌握虚拟机快照的创建方法。
(2)熟练掌握通过快照还原虚拟机系统。

3.2.2 实验环境

(1)硬件环境:计算节点(已安装KVM相关环境)1台,管理节点1台,存储节点1台。
(2)软件环境:virt-manager。

3.2.3 实验原理

snapshot(快照)功能是指可以把虚拟机某个时间点的内存、磁盘文件等的状态保存为一个镜像文件。通过这个镜像文件,可以在以后的任何时间将虚拟机恢复到创建快照时的状态。

当虚拟机处于运行状态时,快照会保存虚拟机当前运行的内存以及磁盘的状态。当用快照恢复虚拟机时,虚拟机可以恢复到快照创建时虚拟机的运行状态;当虚拟机处于关机状态时,快照只保存虚拟机的磁盘镜像状态。

3.2.4 实验内容

本节实验主要有如下内容:①环境准备;②创建虚拟机快照;③利用快照还原系统。

3.2.5 实验步骤

一、环境准备

(1)搭建好 KVM 宿主机。
(2)在 KVM 宿主机上创建 1 台虚拟机。

二、创建/恢复虚拟机快照

(1)在虚拟系统管理器中选中一台虚拟机,点击鼠标右键,选择"打开",打开该虚拟机的控制台(图 3-12)。

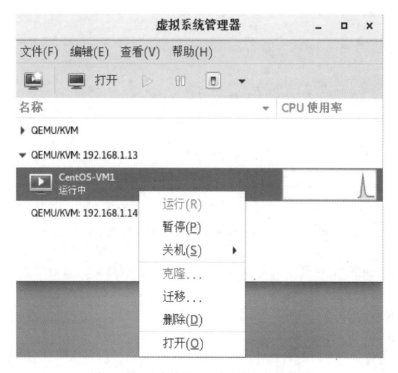

图 3-12 选中并打开一台虚拟机的控制台

(2)点击右上角的"管理虚拟机快照"按钮(图 3-13)。

图 3-13　启动虚拟机的"快照管理器"

(3)点击左下角的"创建新快照"按钮(图 3-14)。

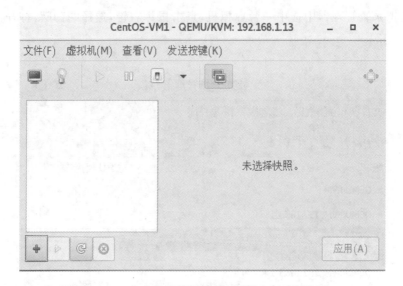

图 3-14　在虚拟机的"快照管理器"中创建新快照

(4)在"创建快照"对话框中填写虚拟机快照的名称及描述信息,然后点击"完成"按钮(图 3-15)。

(5)系统显示已经成功创建了一个虚拟机快照(图 3-16)。

(6)将虚拟机恢复到快照的状态。

当虚拟机持续运行一段时间以后,可以将虚拟机恢复到以前存储的某个快照的状态。通常需要恢复到某个快照的场景是:对虚拟机做了某些复杂的配置或修改后,导致虚拟机崩溃或无法正常工作,此时,希望将虚拟机的系统恢复到对其配置或修改之前正常工作的状态。当虚拟机恢复到快照时,虚拟机当前的运行状态将会被丢弃。

图 3-15　填写虚拟机快照的名称及描述信息

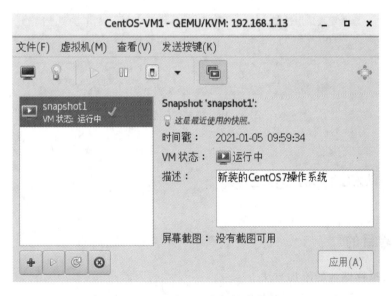

图 3-16　虚拟机快照创建成功

若想将虚拟机恢复到某个快照,只需在该快照上点击右键,然后选择"开始快照"(图 3-17),在系统询问是否确实要应用该快照时,请选择"是"以确定丢弃虚拟机当前状态,并恢复到快照的状态(图 3-18)。

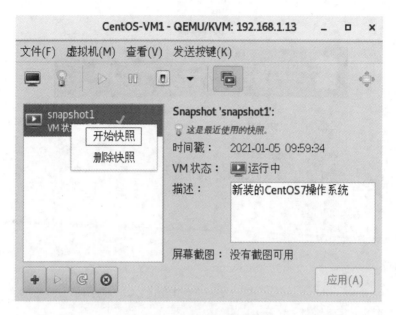

图 3-17　恢复虚拟机的快照

图 3-18　确定恢复快照

3.3　虚拟机的动态迁移

3.3.1　实验目的

（1）掌握共享存储池的添加方法。
（2）掌握虚拟机动态迁移的原理。
（3）掌握虚拟机动态迁移的操作方法。

3.3.2　实验环境

（1）硬件环境：计算节点（已安装 KVM 相关环境）1 台，管理节点 1 台，存储节点 1 台。
（2）软件环境：virt-manager。

3.3.3 实验原理

迁移分为静态迁移(static migration)和动态迁移(live migration)。静态迁移和动态迁移最大的区别在于,静态迁移时,在明显的一段时间内虚拟机中的服务不可用,而动态迁移则没有明显的服务暂停时间。虚拟化环境中的静态迁移分为两种,一种是关闭虚拟机后,将其硬盘镜像复制到另一台宿主机上然后恢复启动,这种迁移不能保留虚拟机中运行的服务;另一种是两台宿主机共享存储系统,也就是两台宿主机的虚拟机都保存在同一个共享存储系统之中,只需要在暂停(而不是完全关闭)虚拟机后复制其内存镜像到另一台宿主机中恢复启动,这种迁移可以保持虚拟机迁移前的内存状态和系统运行的服务。

动态迁移是指在保证虚拟机上的服务正常运行的同时,让虚拟机在不同的宿主机之间进行迁移,其步骤与静态迁移一致,既有磁盘镜像和内存都复制的动态迁移,也有仅仅复制内存镜像的动态迁移。不同的是,为了保证迁移过程中虚拟机服务的可用性,动态迁移必须要求虚拟机在迁移过程中仅有非常短暂的停机时间。

为了使得动态迁移具有高效性,建议虚拟化的环境满足如下要求。

(1)源宿主机和目的宿主机之间使用网络共享的存储系统来保存虚拟机的磁盘镜像。

(2)为了提高动态迁移的成功率,尽量在同类型 CPU 的宿主机上进行动态迁移。

(3)64 位虚拟机只能在 64 位宿主机之间迁移,而 32 位虚拟机可以在 32 位和 64 位宿主机之间迁移。

(4)在进行动态迁移时,被迁移虚拟机的名称必须唯一,在目的宿主机上不能有与源宿主机中被迁移虚拟机同名的虚拟机存在(图 3-19)。

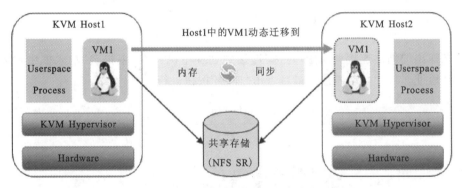

图 3-19 基于共享存储的 KVM 动态迁移

根据动态迁移的要求,为了实现虚拟机的动态迁移必须要做如下几项准备工作:①添加新的 KVM 宿主机,实现动态迁移,至少应有两台一样的 KVM 宿主机;②给每一台 KVM 宿主机添加同一个共享存储池;③在共享存储池中安装虚拟机。

3.3.4 实验内容

本节实验中主要为以下几项内容:①添加新的 KVM 宿主机;②添加共享存储池;③在共享存储池中安装虚拟机;④虚拟机动态迁移。

3.3.5 实验步骤

一、添加新的 KVM 宿主机

在前面的章节中，virt-manager 虚拟机管理软件中已经添加了 1 台 KVM 宿主机，由于实现虚拟机的动态迁移必须至少有 2 台 KVM 宿主机，所以还需要添加 1 台 KVM 宿主机。下面是添加新的 KVM 宿主机的详细步骤。

(1) 在计算节点 A 上安装、配置 KVM 相关环境，并将该 KVM 宿主机名改为 kvm2，IP 地址配置为"192.168.1.14"（kvm1 的 IP 地址设置为"192.168.1.13"）。

(2) 在 virt-manager 虚拟机管理软件的菜单栏点击"文件→添加连接..."（图 3-20）。

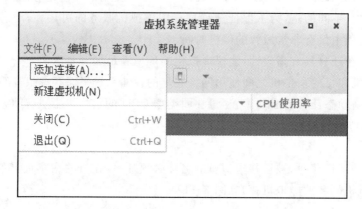

图 3-20　添加新的 KVM 宿主机

(3) 在弹出的创建连接页面中，勾选"连接到远程主机"，在"主机名"输入框中输入 kvm2 主机的 IP 地址，勾选"自动连接"，最后点击"连接"按钮进行连接（图 3-21）。

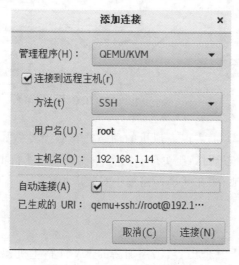

图 3-21　输入主机的 IP 地址

(4)在"OpenSSH"对话框中输入 kvm2 主机的 root 密码,点击"确定"按钮进行连接(图3-22)。

图 3-22 输入主机 kvm2 的 root 用户密码

(5)kvm2 主机添加成功后,在 virt-manager 显示界面中可以看到 kvm2 出现在主机列表中,如图 3-23 所示。

图 3-23 成功添加 KVM 宿主机

二、添加共享存储池

为了实现虚拟机动态迁移的高效性,就必须为 KVM 宿主机添加共享存储池,将 KVM 宿主机所有运行的虚拟机的磁盘镜像保存在共享存储池中。

KVM 也支持多种共享存储类型,包括 NFS、iSCSI、Fiber Channel 等。在本书中采用 NFS 作为 KVM 的共享存储类型,选择该类型共享存储首先需要配置一台 NFS 服务器,然后在该 NFS 服务器上创建一个存储虚拟机磁盘映像的目录,并将该目录的网络权限打开。

本实验中,在存储节点已搭建好 NFS 服务并在上面创建共享目录"/export/nfs_sr",存储节点的 IP 地址为"192.168.1.14",在 KVM 宿主机上添加共享存储的步骤如下。

(1)选中 KVM 宿主机,点击鼠标右键选择"详情"(图 3-24)。

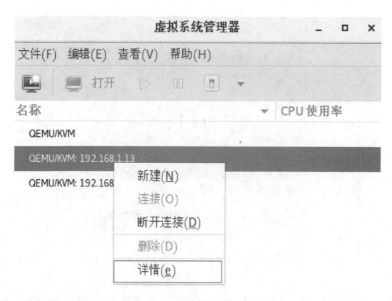

图 3-24　进入 KVM 宿主机的硬件信息页面

注意:用户若是已经创建虚拟机,则应先将该宿主机上的虚拟机全部删除,或者是将该 KVM 宿主机的环境重新安装,然后继续后续步骤。

(2)在弹出的页面中,进入"存储"选项页面,点击左下角的"＋"(添加池)按钮进入添加存储池页面(图 3-25)。

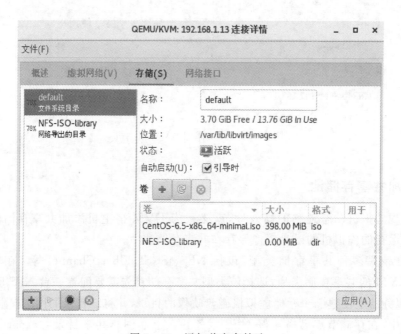

图 3-25　添加共享存储池

(3)输入共享存储池的名称,如"NFS-VirtualDiskStorage",选择存储池的类型"netfs",然后点击"前进"进入下一步(图 3-26)。

图 3-26 输入存储池的名称,选择存储类型

(4)在主机名输入框中输入已经搭建好的 NFS 服务器的 IP 地址,如"192.168.1.14",在源路径输入框中输入 NFS 共享文件夹的路径"/export/nfs_sr"。然后,点击"完成"按钮完成共享存储池的添加(图 3-27)。

图 3-27 输入共享存储池的连接信息

(5)NFS 共享存储池添加完成后,如图 3-28 所示,在该界面的右侧会显示共享存储池的可用空间等信息。

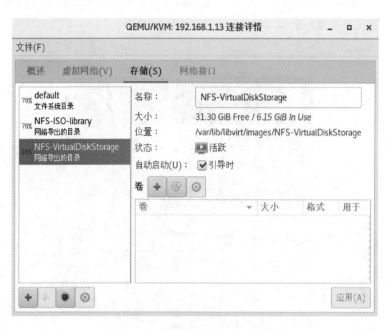

图 3-28 共享存储池成功添加

(6)重新按照(1)~(5)步骤将该共享存储池添加到另一台 kvm2 主机上。

注意：所有的 KVM 宿主机在添加共享存储池时，需保持共享存储池的名称一致，否则实施虚拟机动态迁移时会失败。

三、在共享存储池中安装虚拟机

共享存储池添加好了之后，就可以将虚拟机在共享存储池中安装了，也就是将 KVM 宿主机上运行的虚拟机的磁盘镜像存储在共享存储池中。下面介绍在共享存储池中安装虚拟机的步骤。

(1)在 virt-manager 管理界面中点击菜单栏"文件"下面的"新建虚拟机"图标创建虚拟机，如图 3-29 所示。

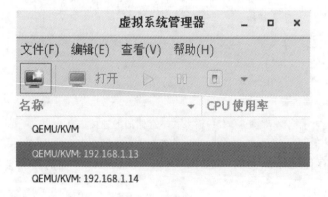

图 3-29 新建虚拟机

(2)在"连接"下拉框中,选择虚拟机运行的宿主机(此处选择 kvm1 主机,即选择 kvm1 对应的 IP 地址),勾选"本地安装介质",点击"前进"按钮进入下一步(图 3-30)。

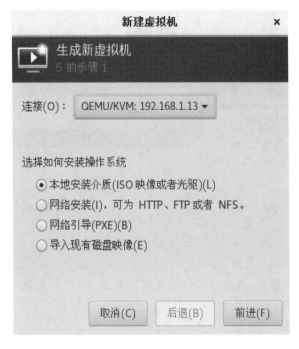

图 3-30 输入虚拟机名称,选择虚拟机的宿主机

(3)选择虚拟机安装所需的系统 ISO 镜像。
①勾选"使用 ISO 映像",点击"浏览"按钮选择 ISO 镜像(图 3-31)。

图 3-31 选择 ISO 镜像(一)

②选中"NFS-ISO-library",然后在右侧页面中选中所需的系统镜像,此处选择"CentOS-6.5-mini.iso",选中后点击"选择卷"(图 3-32)。

图 3-32　选择 ISO 镜像(二)

③选好系统 ISO 镜像后,再选择操作系统类型和版本,然后点击"前进"按钮进入下一步(图 3-33)。

图 3-33　完成 ISO 镜像的选择

注意:如果选择 kvm2 主机作为虚拟机的宿主机,需先将 ISO 存储池添加到 kvm2 主机上,否则在系统 ISO 做镜像选择时,将没有 ISO 镜像可以选择。

(4)配置虚拟机的 CPU、内存,点击"前进"进入下一步(图 3-34)。

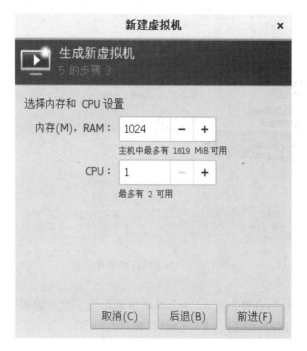

图 3-34 设置虚拟机的 CPU、内存

(5)选择虚拟机的磁盘镜像存储位置。

①勾选"选择或创建自定义存储",点击"管理..."按钮定义虚拟机的存储路径(图 3-35)。

图 3-35 定义虚拟机的存储位置

② 在存储池中选中共享存储池,然后点击"新加存储卷"为虚拟机创建存储卷(图 3-36)。

图 3-36　为虚拟机创建存储卷

③ 选择磁盘卷的格式,选择"qcow2"(选择此格式是为了支持虚拟机的快照功能),设置磁盘卷的大小,此处选择 8GiB,然后点击"完成"按钮(图 3-37)。

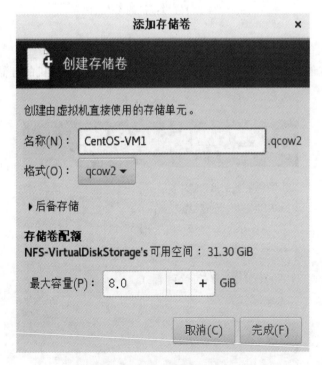

图 3-37　设置磁盘卷的大小

④ 选中刚创建的存储卷,然后点击"选择卷"(图 3-38)。

图3-38 选择新创建的存储卷

⑤选好虚拟机的磁盘镜像后,点击"前进"按钮进入下一步(3-39)。

图3-39 完成虚拟机存储卷的选择

(6)点击"完成"开始虚拟机的创建(图3-40)。

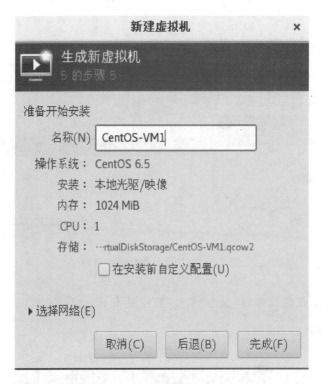

图 3-40 完成虚拟机的配置

(7)虚拟机创建完成后,virt-manager 自动打开该虚拟机的控制台界面,虚拟机进入操作系统的安装界面(图 3-41)。

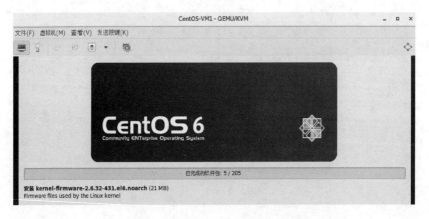

图 3-41 虚拟机开始安装系统

说明:Linux 虚拟机安装完成后仍需要安装 ACPI,否则在 virt-manager 中无法对虚拟机实行关机、重启操作。安装方法如下。

```
[root@localhost ~]# yum install -y acpid
[root@localhost ~]# service acpid start
[root@localhost ~]# chkconfig acpid on
```

四、虚拟机动态迁移

上述准备工作做完后,就可以进行虚拟机动态迁移的操作了。如图 3-42 所示,在 kvm1 (192.168.1.13)主机上运行了一台 CentOS 系统的虚拟机,在 kvm2(192.168.1.14)主机上没有运行虚拟机。下面介绍将 kvm1 主机上运行的虚拟机动态迁移到 kvm2 主机上的详细步骤。

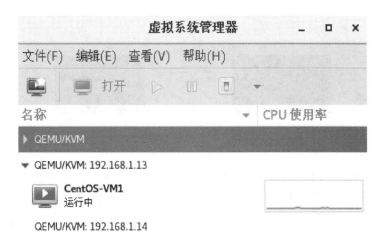

图 3-42 虚拟机、主机列表

(1)为了检测虚拟机在动态迁移的过程中是否出现过服务中断,可以做出如下测试,在管理节点上 ping 该虚拟机的 IP 地址,假如该虚拟机的 IP 地址为"192.168.1.4",ping 命令的结果如下(图 3-43)。

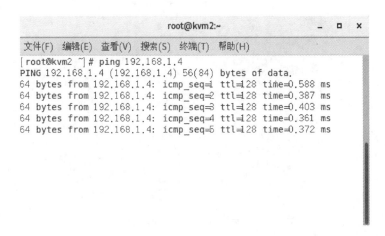

图 3-43 ping 虚拟机

(2)选中虚拟机,点击鼠标右键选择"迁移..."(图 3-44)。

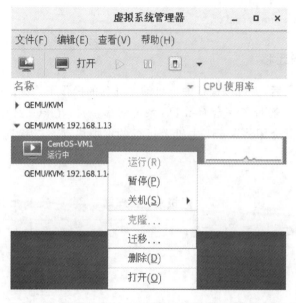

图 3-44 迁移虚拟机

(3)在弹出的页面中,选择新主机为 kvm2,模式为"直连",端口设置为"让 libvirt 自动判断",在高级选项中勾选"允许不可靠",然后点击"迁移"按钮,开始虚拟机的动态迁移(图 3-45)。

图 3-45 选择虚拟机迁移的目的主机

(4)虚拟机动态迁移的进度显示(图 3-46)。

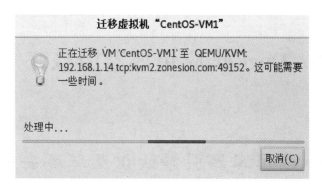

图 3-46 虚拟机迁移进度

(5)虚拟机在动态迁移过程中,继续观察步骤(1)中对虚拟机的 ping 结果,如图 3-47 所示,可以发现虚拟机一直可以 ping 通,说明虚拟机的服务在迁移过程中一直正常运行。

图 3-47 检测被迁移虚拟机上的服务是否停止

(6)虚拟机迁移结束后,在 virt-manager 管理界面中可以看到虚拟机 CentOS-VM1 已经从 KVM1 主机迁移到 KVM2 主机上,并且在 KVM2 主机上继续保持运行状态(图 3-48)。

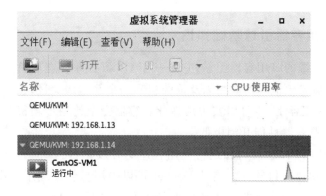

图 3-48 虚拟机迁移成功

4 Docker 应用容器引擎开发

4.1 Docker 应用容器引擎认识及环境安装实验

4.1.1 实验目的

(1) 认识 Docker 应用容器引擎。
(2) 安装 Docker 引擎。
(3) 配置 Docker 服务。

4.1.2 实验环境

(1) 硬件：PC 机。
(2) 软件：CentOS 7 及以上系统、Windows 7 及以上系统。
(3) 正常互联网网络连接。

4.1.3 实验内容

(1) 认真阅读基础知识内容，总结 Docker 应用容器引擎。
(2) 按照提示动手安装 Docker 应用容器引擎。

4.1.4 实验步骤

一、Docker 应用容器引擎基础知识

Docker 是一个开源的应用容器引擎，它是基于 Go 语言并且遵从 Apache2.0 协议开源。Docker 可以让开发者打包他们的应用以及依赖包到一个轻量级、可移植的容器中，然后发布到任何流行的 Linux 机器上，也可以实现虚拟化。容器完全使用沙箱机制，相互之间不会有任何接口，更重要的是容器性能开销极低。

Docker 是一个用于开发、交付和运行应用程序的开放平台。Docker 将应用程序与基础架构分开，从而可以快速交付软件。借助 Docker 可以采用与管理应用程序相同的方式来管理基础架构。通过利用 Docker 的方法来快速交付、测试和部署代码，从而大大减少编写代码和在生产环境中运行代码之间的延迟，Docker 具有以下优点。

1. 快速一致地交付应用程序

Docker 允许开发人员使用提供的应用程序或服务的本地容器在标准化环境中工作,从而简化了开发的生命周期。

容器非常适合持续集成和持续交付(CI/CD)工作流程,示例如下:

开发人员在本地编写代码,并使用 Docker 容器与同事共享他们的工作。

他们使用 Docker 将其应用程序推送到测试环境中,并执行自动或手动测试。

当开发人员发现错误时,可以在开发环境中对其进行修复,然后将其重新部署到测试环境中,以进行测试和验证。

测试完成后,将修补程序推给生产环境,就像将更新的镜像推送到生产环境一样简单。

2. 响应式部署和扩展

Docker 是基于容器的平台,允许高度可移植的工作负载。Docker 容器可以在开发人员的本机上、数据中心的物理或虚拟机上、云服务上或混合环境中运行。

Docker 的可移植性和轻量级的特性,还可以使开发人员轻松地完成动态管理的工作,并根据业务需求指示,实时扩展或拆除应用程序和服务。

3. 在同一硬件上运行更多工作负载

Docker 轻巧快速,它为基于虚拟机管理程序的虚拟机提供了可行、经济、高效的替代方案。因此,开发人员可以利用更多的计算能力来实现业务目标。Docker 非常适合于高密度环境以及中小型部署,而开发人员可以用更少的资源做更多的事情。

二、Docker 安装

1. CentOS Docker 安装

Docker 支持 64 位 CentOS 7 及以上版本安装,可以通过下列方式安装。

1)自动安装

如果使用官方安装脚本自动安装,安装命令如下:

```
curl -fsSL https://get.docker.com | bash -s docker --mirror Aliyun
```

也可以使用国内 daocloud 一键安装命令:

```
curl -sSL https://get.daocloud.io/docker | sh
```

2)手动安装

如果使用手动安装,步骤如下:

(1)卸载旧版本。较旧的 Docker 版本称为 docker 或 docker-engine。如果已安装这些旧程序,请卸载它们以及相关的依赖项。

```
$ sudo yum remove docker \
          docker-client \
          docker-client-latest \
          docker-common \
          docker-latest \
          docker-latest-logrotate \
          docker-logrotate \
          docker-engine
```

（2）安装 Docker Engine-Community。使用 Docker 仓库进行安装，在新主机上首次安装 Docker Engine-Community 之前，需要设置 Docker 仓库，然后可以从仓库安装和更新 Docker。

（3）设置仓库。安装所需的软件包。yum-utils 提供了 yum-config-manager，并且 device mapper 存储驱动程序需要 device-mapper-persistent-data 和 lvm2。

```
$ sudo yum install -y yum-utils \
  device-mapper-persistent-data \
  lvm2
```

可以使用以下命令来设置稳定的仓库。可以使用官方源地址，但是比较慢。

```
$ sudo yum-config-manager \
    --add-repo \
    https://download.docker.com/linux/centos/docker-ce.repo
```

也可以选择国内的一些源地址，如阿里云或清华大学源。

```
$ sudo yum-config-manager \
    --add-repo \
    http://mirrors.aliyun.com/docker-ce/linux/centos/docker-ce.repo
$ sudo yum-config-manager \
    --add-repo \
    https://mirrors.tuna.tsinghua.edu.cn/docker-ce/linux/centos/docker-ce.repo
```

（4）安装最新版本的 Docker Engine-Community 和 containerd，或者转到下一步安装特定版。

```
$ sudo yum install docker-ce docker-ce-cli containerd.io
```

如果提示是否接受 GPG 密钥，请选择"是"。

Docker 安装完默认未启动，并且已经创建好 docker 用户组，但该用户组下没有用户。

如果启用了多个 Docker 仓库，而且未在 yum install 或 yum update 命令中指定版本，将安装或更新至最高版本，但最高版本可能不太稳定。

要安装特定版本的 Docker Engine-Community，需在存储库中列出可用版本，然后选择并安装。

①列出并排序您存储库中可用的版本。此示例按版本号（从高到低）对结果进行排序。

```
$ yum list docker-ce --showduplicates | sort -r
docker-ce.x86_64    3:18.09.1-3.el7         docker-ce-stable
docker-ce.x86_64    3:18.09.0-3.el7         docker-ce-stable
docker-ce.x86_64    18.06.1.ce-3.el7        docker-ce-stable
docker-ce.x86_64    18.06.0.ce-3.el7        docker-ce-stable
```

②通过其完整的软件包名称安装特定版本,该软件包名称是软件名(docker-ce)加上版本字符串(第二列),中间用连字符(-)分隔,如 docker-ce-18.09.1。

```
$ sudo yum install docker-ce-<VERSION_STRING> docker-ce-cli-<VERSION_STRING> containerd.io
```

③启动 Docker。

```
$ sudo systemctl start docker
```

④通过运行 hello-world 映像来验证是否正确安装了 Docker Engine-Community。

```
$ sudo docker run hello-world
```

2. Windows Docker 安装

Windows 7、Windows 8、Windows 10 等系统,需要利用 docker toolbox 来安装,国内可以使用阿里云的镜像来下载,下载地址为"http://mirrors.aliyun.com/docker-toolbox/windows/docker-toolbox/"。安装比较简单,双击"运行",点击"Next"即可,可以勾选自己需要的组件,如图 4-1 所示。

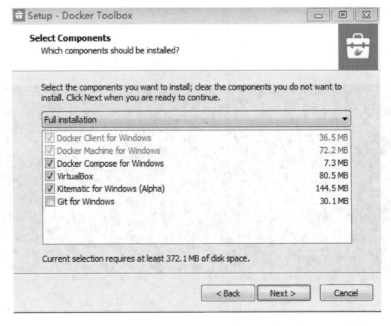

图 4-1　docker toolbox 安装图

docker toolbox 是一个工具集,它主要包含以下内容。
(1) Docker CLI——客户端,用来运行 docker 引擎创建镜像和容器。
(2) Docker Machine——可以在 Windows 的命令行中运行 docker 引擎命令。
(3) Docker Compose——用来运行 docker-compose 命令。
(4) Kitematic——这是 Docker 的 GUI 版本。
(5) Docker Quickstart shell——这是一个已经配置好 Docker 的命令行环境。
(6) Oracle VM Virtualbox——虚拟机。

下载完成之后直接点击"安装",安装成功后,桌边会出现 3 个图标,如图 4－2 所示。

图 4－2　docker toolbox 安装完成示意图

点击 Docker Quickstart 图标来启动 Docker Toolbox 终端。

如果系统显示 User Account Control 窗口来运行 VirtualBox 修改电脑,选择"Yes",如图 4－3 所示。

图 4－3　Docker Toolbox 终端启动图

$符号后面可以输入以下命令来执行。

```
$ docker run hello-world
Unable to find image 'hello-world:latest' locally
Pulling repository hello-world
91c95931e552：Download complete
a8219747be10：Download complete
Status：Downloaded newer image for hello-world:latest
Hello from Docker.
This message shows that your installation appears to be working correctly.

To generate this message，Docker took the following steps：
1. The Docker Engine CLI client contacted the Docker Engine daemon.
2. The Docker Engine daemon pulled the "hello-world" image from the Docker Hub.
   (Assuming it was not already locally available.)
3. The Docker Engine daemon created a new container from that image which runs the
   executable that produces the output you are currently reading.
4. The Docker Engine daemon streamed that output to the Docker Engine CLI client，which sent it
   to your terminal.
To try something more ambitious，you can run an Ubuntu container with：
$ docker run -it ubuntu bash
For more examples and ideas，visit：
https://docs.docker.com/userguide/
```

对于 Windows 10 系统，Docker 有专门的 Windows 10 专业版系统的安装包，需要开启 Hyper-V。首先，开启 Hyper-V，如图 4-4 所示。在 Windows 10 的启用或关闭 Windows 功能中选中"Hyper-V"。

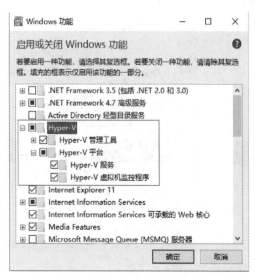

图 4-4 开启"Hyper-V"

(1)安装 Toolbox。最新版 Toolbox 下载地址为"https://www.docker.com/get-started",注册 1 个账号,然后登录。点击"Get started with Docker Desktop",并下载 Windows 的版本(图 4-5),如果用户还没有登录,会被要求注册登录。

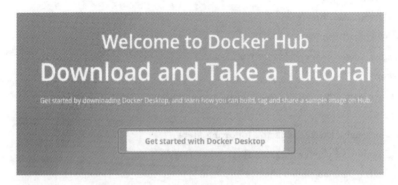

图 4-5 下载 Toolbox

也可以不用注册,直接下载 Windows 版(图 4-6)。

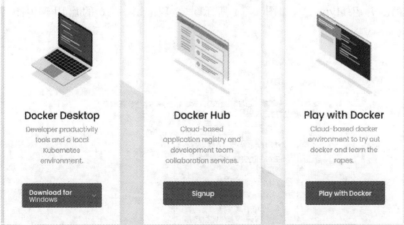

图 4-6 下载 Toolbox Windows 版

(2)运行安装文件。双击下载的 Docker for Windows Installer 的安装文件,一路点击"Next",最后点击"Finish"完成安装(图 4-7)。

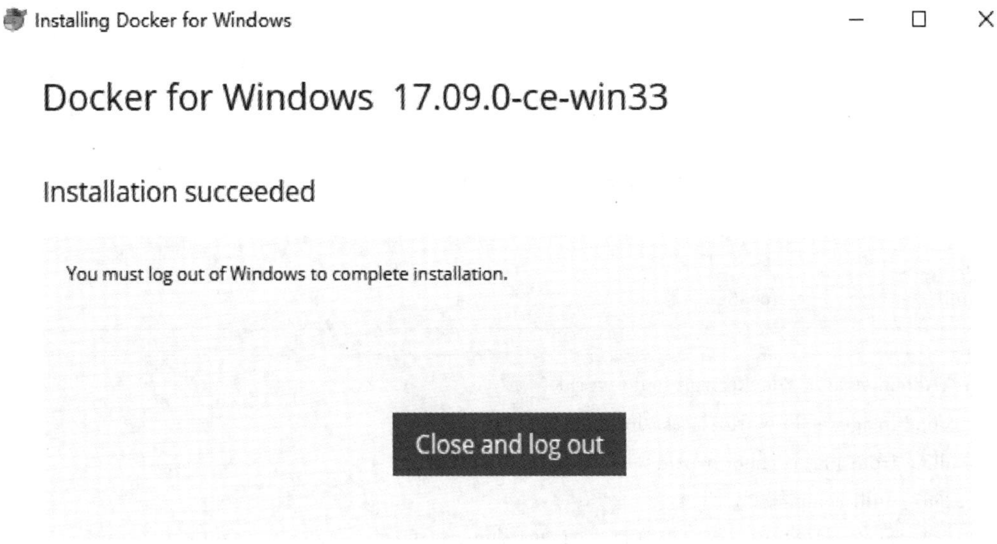

图 4 - 7 安装 Toolbox

安装完成后，Docker 会自动启动。通知栏上会出现一个类似小鲸鱼的图标，这表示 Docker 正在运行。可以在命令行执行 docker version 来查看版本号，执行 docker run helloworld 来载入测试镜像测试。

```
C：\Users\Administrator>docker version
Client：Docker Engine-Community
Azure integration       0.1.15
version：                19.03.12
API version ：           1.40
Go version：             go1.13.10
Git commit：             48a66213fe
Built：                  Mon Jun 22 15：43：18 2020
OS/Arch：                windows/amd64
Experimental：           false
Server ：Docker Engine-Community
Engine：
Version ：               19.03.12
API version ：           1.40(minimum version 1.12)
Go version：             go1.13.10
Git commit：             48a66213fe
Built：                  Mon Jun 22 15：49：27 2020
OS/Arch：                linux/amd64
Experimental：           false
```

```
containerd：
version：              v1.2.13
GitCommit：            7ad184331fa3e55e52b890ea95e65ba581ae3429
runc：
version：              1.0.0-rc10
GitCommit：            dc9208a3303feef5b3839f4323d9beb36dfOa9dd
docker-init：
version：              0.18.0
GitCommit：            fec3683
```

```
C：Users\Administrator>docker run hello-world
Unable to find image 'hello-world：latest' locally
latest：Pulling from library/hello-world
0e03bdcc26d7：Pull complete
Digest：sha256：4cf9c47f86df71d48364001ede3a4fcd85ae80ce02ebad74156906caff5378bc
Status：Downloaded newer image for hello-world：latest
Hello from Docker！
This message shows that your installation appears to be working correctly.
```

如果没启动，可以在 Windows 搜索 Docker 来启动；启动后，也可以在通知栏上看到 图标。

（3）镜像加速。对于使用 Windows 10 的系统，在系统右下角托盘"Docker"图标内右键菜单选择"Settings"，打开配置窗口后左侧导航菜单选择"Docker Engine"。在"Registry-mirrors"一栏中填写加速器地址"https：//registry.docker-cn.com"，点击"Apply&Restart"保存并重启，Docker 就会重启并应用配置的镜像地址了（图 4-8）。

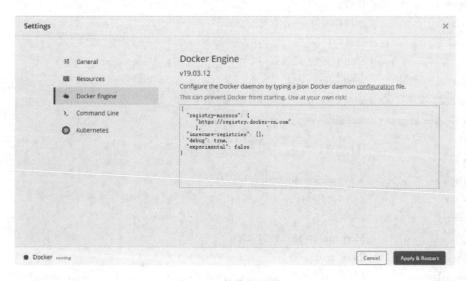

图 4-8　镜像加速设置

4.2 Docker 基础操作实验

4.2.1 实验目的

(1)熟悉 Docker 容器使用。
(2)熟悉 Docker 镜像使用。
(3)熟悉 Docker 容器连接。

4.2.2 实验环境

(1)硬件:PC 机。
(2)软件:CentOS 7 及以上系统、Windows 7 及以上系统。
(3)正常互联网网络连接。

4.2.3 实验内容

(1)Docker 容器使用。
(2)Docker 镜像使用。
(3)Docker 容器连接。

4.2.4 实验步骤

一、Docker 容器使用

1. 启动查看 Docker 客户端

启动查看 Docker 客户端非常简单,可以直接输入 docker 命令来查看 Docker 客户端的所有命令。

```
[root@localhost heathcliff] # docker
Usage:docker [OPTIONS] COMMAND
A self-sufficient runtime for containers
options:
  --config string        Location of client config files (default "/root/.docker")
  -c,--context string    Name of the context to use to connect to the daemon (overrides DOCKER_HOST
                         env var and default context set with "docker context use")
  -D,--debug             Enable debug mode
  -H,--host list         Daemon socket(s) to connect to
  -l,--log-level string  Set the logging level ("debug"|"info"|"warn"|"error"|"fatal") (default "info")
```

--tls	Use TLS; implied by . --tlsverify
--tlscacert string	Trust certs signed only by this CA (default "/root/.docker/ca.pem")
--tlscert string	Path to TLS certificate file (default "/root/.docker/cert.pem")
--tlskey string	Path to TLS key file (default "/root/.docker/key.pem")
--tlsverify	Use TLS and verify the remote
-v, --version	Print version information and quit

可以通过命令 docker command --help 更深入地了解指定的 Docker 命令使用方法。

例如要查看 docker stats 指令的具体使用方法如下。

```
[root@localhost heathcliff] # docker stats --help
Usage: docker stats [OPTIONS] [CONTAINER...]
Display a live stream of container(s) resource usage statistics
options:
-a, --all          Show all containers (default shows just running)
    --format string  Pretty-print images using a Go template
    --no-stream     Disable streaming stats and only pull the first result
    --no-trunc      Do not truncate output
```

2. 容器使用

(1) 获取镜像。如果本地没有 ubuntu 镜像，可以使用 docker pull 命令来载入 ubuntu 镜像。

```
[root@localhost heathcliff] #docker pull ubuntu
using default tag: latest
latest: Pulling from library/ubuntu
d72e567cc804: Pull complete
0f3630e5ff08: Pull complete
b6a83d81dlf4: Pull complete
Digest: sha256: bc2f7250f69267c9c6b66d7b6a81a54d3878bb85f1ebb5f951c896d13e6ba537
Status: Downloaded newer image for ubuntu: latest
docker.io/library/ubuntu: latest
```

(2) 启动容器。使用 ubuntu 镜像启动一个容器，参数为以命令行模式进入该容器。

```
docker run -it ubuntu /bin/bash
```

参数说明：

-i：交互式操作。

-t：终端。

ubuntu：ubuntu 镜像。

/bin/bash：放在镜像名后的是命令，这里希望有个交互式 Shell，因此用的是 /bin/bash。

```
[root@localhost heathcliff]# docker run -it ubuntu /bin/bash
root@4ad8f7f118c5:/#
```

(3) 如果要退出终端,直接输入 exit 命令。

```
[root@localhost heathcliff]# docker run -it ubuntu /bin/bash
root@4ad8f7f118c5:/# exit
exit
```

(4) 启动已停止运行的容器。查看所有的容器命令如下:

```
[root@localhost heathcliff]# docker ps -a
CONTAINER ID    IMAGE         COMMAND       CREATED          STATUS              PORTS    NAMES
4ad8f7f118c5    ubuntu        "/bin/bash"   4 minutes ago    Exited (0) 3 minutes ago     agitated_khayyam
bcd27fbacael    ubuntu        "/bin/bash"   14 ninutes ago   Exited (0) 4 minutes ago     elastic_Jepsen
043651b92bc     hello-world   "/hello"      13 hours ago     Exited (0) 13 hours ago      amazing_kapitsa
```

使用 docker start 启动一个已停止的容器,命令如下:

```
docker start <容器 ID>
```

例如:

```
[root@localhost heathcliff]# docker start 4ad8f7f118c5
4ad8f7f118c5
```

(5) 后台运行。在大部分的场景下,操作人员希望 docker 的服务是在后台运行的,可以通过-d 指定容器的后台运行模式。

```
[root@localhost heathcliff]# docker run -itd--name ubuntu-test ubuntu /bin/bash
7695e5f0d0987e438ca8f60cd540663d4 b84859edc6bcd5d20f37c0bc03e15ac
```

```
[root@localhost heathcliff]# docker ps
CONTAINER ID    IMAGE     COMMAND       CREATED          STATUS          PORTS    NAMES
7695e5 f0d098   ubuntu    "/bin/bash"   46 seconds ago   Up 45 seconds            ubuntu-test
```

注:加了-d 参数默认不会进入容器,想要进入容器需要使用指令 docker exec(下面会介绍到)。
(6) 停止一个容器。停止容器的命令如下:

```
docker stop <容器 ID>
```

如:

```
[root@localhost heathcliff]# docker stop 7695e5f0d098
7695e5f0d098
```

停止的容器可以通过 docker restart 重启:

```
docker restart <容器 ID>
```

如:

```
[root@localhost heathcliff]# docker restart 7695e5f0d098
7695e5f0d098
```

(7)进入容器。在使用-d参数时,容器启动后会进入后台。此时想要进入容器,可以通过以下指令:"docker attach"或"docker exec"。推荐使用 docker exec 命令,因为此命令退出容器终端,不会导致容器的停止。

docker attach 命令格式如下:

```
docker attach <容器 ID>
```

如:

```
[root@localhost heathcliff]# docker attach 7695e5f0d098
root@7695e5f0d098:/# exit
exit
[root@localhost heathliff]# docker ps
CONTAINER ID    IMAGE    COMMAND      CREATED         STATUS          PORTS    NAMES
4ad8f7f118c5    ubuntu   "/bin/bash"  17 minutes ago  Up 10 minutes            agitated_khayyam
```

注意:如果从这个容器退出,会导致容器的停止。

docker exec 命令格式如下:

```
docker exec-it <容器 ID> /bin/bash
```

如:

```
[root@localhost heathcliff]# docker exec-it 7695e5f0d098 /bin/bash
root@7695e5f0d098:/# exit
exit
[root@localhost heathcliff]# docker ps
CONTAINER ID    IMAGE    COMMAND      CREATED         STATUS          PORTS    NAMES
7695e5 f0d098   ubuntu   "/bin/bash"  11 minutes ago  Up 17 seconds            ubuntu-test
4ad8f7f118c5    ubuntu   "/bin/bash"  20 minutes ago  Up 13 minutes            agitated_khayyam
```

注意:如果从这个容器退出,不会导致容器的停止,这就是推荐使用 docker exec 的原因。更多参数说明请使用 docker exec--help 命令查看。

(8)导出和导入容器。如果要导出本地某个容器,可以使用 docker export 命令,格式如下:

```
docker export  <容器 ID>   > ubuntu.tar
```

如:导出容器 7695e5f0d098 快照到本地文件 ubuntu.tar。

```
[root@localhost heathcliff]# docker ps
CONTAINER ID    IMAGE    COMMAND      CREATED         STATUS         PORTS    NAMES
7695e5f0d098    ubuntu   "/bin/bash"  16 minutes ago  Up 4 minutes            ubuntu-test
[root@localhost heathcliff]# docker export 7695e5f0d098 >ubuntu.tar
```

这样将导出容器快照到本地文件(图 4-9)。

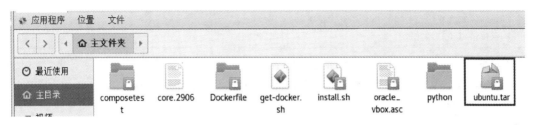

图 4-9 导出的容器快照文件

可以使用 docker import 从容器快照文件中再导入为镜像,以下实例将快照文件 ubuntu.tar 导入到镜像 test/ubuntu v1(注意:cat 后面的路径为 ubuntu.tar 存放的路径)。

```
cat /home/heathcliff/ubuntu.tar | docker import - test/ubuntu:v1
```

如:

```
[root@localhost heathcliff]# cat /home/heathcliff/ubuntu.tar | docker import - test/ubuntu:v1
sha256:ee1181d039b3d9e5f74a7817aa4b1e1c169e4899d764efb356bc37ad5865fc10
[root@localhost heathcliff]# docker images
REPOSITORY      TAG        IMAGE ID         CREATED          SIZE
test/ubuntu     v1         ee1181d039b3     13 seconds ago   72.9MB
```

此外,也可以通过指定 URL 或者某个目录来导入,例如:

```
docker import http://example.com/exampleimage.tgz example/imagerepo
```

(9)删除容器。删除容器使用 docker rm 命令,格式如下:

```
docker rm -f <容器 ID>
```

如:

```
[root@localhost heathcliff]# docker ps
CONTAINER ID   IMAGE     COMMAND       CREATED            STATUS             PORTS    NAMES
7695e5f0d098   ubuntu    "/bin/bash"   About an hour ago  Up About an hour            ubuntu test
[root@localhost heathcliff]# docker rm -f 7695e5f0d098
7695e5f0d098
[root@localhost heathcliff]# docker ps
CONTAINER ID   IMAGE     COMMAND       CREATED            STATUS             PORTS    NAMES
```

下面的命令可以清理掉所有处于终止状态的容器。

```
docker container prune
```

如:

```
[root@localhost heathcliff]# docker container prune
WARNING! This will remove all stopped containers.
Are you sure you want to continue? [y/N] n
Total reclaimed space:0B
```

(10)运行一个 Web 应用。前面运行的容器并没有一些特别的用处。接下来尝试使用 docker 构建一个 Web 应用程序。我们将在 docker 容器中运行一个 Python Flask 应用来运行一个 Web 应用。

```
docker pull training/webapp    #载入镜像
docker run -d -P training/webapp python app.py
```

参数说明：
-d:让容器在后台运行。
-P:将容器内部使用的网络端口随机映射到使用的主机上。

```
[root@localhost heathcliff]# docker pull training/webapp
using default tag:latest
latest:Pulling from training/webapp
Image docker.io/training/webapp:latest uses outdated schemal manifest format. Please upgrade.com/regis-
try/spec/deprecated-schema-v1/
e190868d63f8:Pull complete
909cd34c6fd7:Pull complete
0b9bfabab7c1:Pull complete
a3ed95caeb02:Pull complete
10bbbc0fc0ff:Pull complete
fca59b508e9f:Pull complete
e7ae2541b15b:Pull complete
9dd97ef58ce9:Pull complete
a4c1b0cb7af7:Pull complete
Digest:sha256:06e9cl983bd6d5db5fba376ccd63bfa529e8d02f23d5079b8f74a616308fblld
Status:Downloaded newer image for training/webapp:latest
```

```
[root@localhost heathcliff]# docker run -d -P training/webapp python app.py
f853697 b56 b342 a60b9b44el0f f27e891c785c4d52907c72f8f77e885c2b8a7e
```

(11)查看 Web 应用容器。使用 docker ps 来查看正在运行的容器。

```
[root@localhost heathcliff]# docker ps
CONTAINER ID  IMAGE          COMMAND         CREATED      STATUS     PORTS                   NAMES
f8536976563   training/webap "python app.py" 9 minutes ago Up 9 minutes 0.0.0.0:32768→5000/tcp distracted_johnson
```

这里多了端口信息,PORTS:0.0.0.0:32768→5000/tcp。

Docker 开放了 5000 端口(默认 Python Flask 端口)映射到主机端口 32768 上。

这时可以通过浏览器访问 Web 应用(图 4-10)。

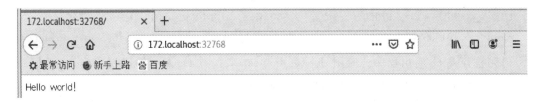

图 4-10 测试 Flask 发布的 Web 应用

也可以通过 -p 参数来设置不一样的端口。

```
docker run -d -p 5000:5000 training/webapp python app.py
```

docker ps 查看正在运行的容器，容器内部的 5000 端口映射到本地主机的 5000 端口上。

```
[root@localhost heathcliff]# docker run -d -p 5000:5000 training/webapp python app.py
2d02ee564ab39330d2952ab957c1b3e995e3fb29d2e4e08bfe133a0b489f5c5a
[root@localhost heathcliff]# docker ps
CONTAINER ID    IMAGE            COMMAND           CREATED          STATUS          PORTS                      NAMES
2d02ee564ab3    training/webapp  "python app.py"   8 seconds ago    Up 6 seconds    0.0.0.0:5000->5000/tcp     optimistic_perlman
f853697b563     training/webapp  "python app.py"   6 hours ago      Up 6 hours      0.0.0.0:32768->5000/tcp    distracted_johnson
```

(12) 网络端口的快捷方式。通过 docker ps 命令可以查看容器的端口映射，docker 还提供了另一个快捷方式：docker port，使用 docker port 可以查看指定（ID 或者名字）容器的某个确定端口映射到宿主机的端口号。

上面创建的 Web 应用容器 ID 为 "2d02ee564ab3" 名字为 "optimistic_perlman"。

可以使用 "docker port 2d02ee564ab3" 或 "docker port optimistic_perlman" 来查看容器端口的映射情况。

```
[root@localhost heathcliff]# docker port 2d02ee564ab3
5000/tcp-> 0.0.0.0:5000
[root@localhost heathcliff]# docker port optimistic_perlman
5000/tcp-> 0.0.0.0:5000
```

(13) 查看 Web 应用程序日志。docker logs（ID 或者名字）可以查看容器内部的标准输出。

```
docker logs -f 2d02ee564ab3
```

-f：让 docker logs 像使用 tail -f 一样来输出容器内部的标准输出。

如：

```
[root@localhost heathcliff]   docker logs -f 2d02ee564ab3
 *  Running on http://0.0.0.0:5000/(Press CTRL+C to quit)
```

从上面可以看到，应用程序使用的是 5000 端口，并且能够查看到应用程序的访问日志。

(14) 查看 Web 应用程序容器的进程。使用 docker top 来查看容器内部运行的进程：

```
[root@localhost heathcliff]# docker top optimistic_perlman
UID    PID    PPID    C    STIME    TTY    TIME      CHD
root   10163  10146   0    16:44    ?      0:00:00   python app.py
```

(15)检查 Web 应用程序。使用 docker inspect 来查看 Docker 的底层信息,会返回一个 JSON 文件记录着 Docker 容器的配置和状态信息。

```
[root@localhost heathcliff]# docker inspect optimistic_perlman
{
  "Id": "2d02ee564ab39330d2952ab957c1b3e995e3fb29d2e4e08bfe133a0b489f5c5a",
  "Created": "2020-09-28T08:44:17.97988267Z",
  "Path": "python",
  "Args":{
  "app.py"
  },
  "State": {
    "Status": " running",
    "Running": true,
    "Paused": false,
    "Restarting": false,
    "00MKilled": false,
    "Dead": false,
    "Pid": 10163,
    "ExitCode": 0,
    "Error": " " ,
    "startedAt":"2020-09-28T08:44:18.5032000892",
    "FinishedAt": "0001-01-01 T00:00:00Z"
  },
  "image": "sha256:6fae60ef344644649a39240b94d73b8ba9c67f898ede85cf8e947a887b3e6557",
  "ResolvConfPath" :"/var/lib/docker/containers/2d02ee564ab39330d2952ab957c1b3e995e3fb29d2e4
```

(16)停止 Web 应用容器。

```
[root@localhost heathcliff]# docker stop optimistic_perlman
optimistic_perlman
```

(17)重启 Web 应用容器。已经停止的容器,可以使用命令 docker start 来启动。

```
[root@localhost heathcliff]# docker start optimistic_perlman
optimistic_perlman
```

查询最后一次创建的容器。

```
[root@localhost heathcliff]# docker ps -l
CONTAINER ID   IMAGE            COMMAND          CREATED       STATUS          PORTS                    NAMES
2d02ee564ab3   training/webapp  "python app.py"  3 hours ago   Up 49 minutes   0.0.0.0:5000->5000/tcp   optimistic_perlman
```

正在运行的容器,可以使用 docker restart 命令来重启。

```
docker restart <容器 ID>
```

如:

```
[root@localhost heathcliff]# docker restart  2d02ee564ab3
2d02ee564ab3
```

(18)移除 Web 应用容器。可以使用 docker rm 命令来删除不需要的容器。

```
[root@localhost heathcliff]# docker rm optimistic_perlman
optimistic_perlman
```

删除容器时,容器必须是停止状态,否则会报如下错误。

```
[root@localhost heathcliff]# docker rm optimistic_perlman
Error response from daemon: You cannot remove a running container 2d02ee564ab39330d2952ab957c1b3e995e3fb29d2e4e08bfe133a0b489f5c5a. Stop the container before attempting removal or force remove
```

二、Docker 镜像使用

当运行容器时,使用的镜像如果在本地不存在,Docker 就会自动从 docker 镜像仓库中下载,默认是从 Docker Hub 公共镜像源下载。

1. 列出镜像列表

可以使用 docker images 来列出本地主机上的镜像。

```
docker images
```

各个选项说明:
REPOSITORY:表示镜像的仓库源。
TAG:镜像的标签。
IMAGE ID:镜像 ID。
CREATED:镜像创建时间。
SIZE:镜像大小。

```
[root@localhost heathcliff]# docker images
REPOSITORY        TAG      IMAGE ID        CREATED       SIZE
ubuntu            latest   91401 08b62dc   2 days ago    72.9MB
hello-world       latest   bf756 fb1 ae65  8 months ago  13.3KB
training/webapp   latest   6fae60ef3446    5 years ago   349MB
```

同一仓库源可以有多个 TAG,代表这个仓库源的不同个版本,如 ubuntu 仓库源有 15.10、14.04 等多个不同的版本,可使用 REPOSITORY:TAG 来定义不同的镜像。

所以,如果要使用版本为 15.10 的 ubuntu 系统镜像来运行容器,命令如下:

```
docker run -t -i ubuntu:15.10 /bin/bash
```

参数说明:

-i:交互式操作。

-t:终端。

ubuntu:15.10:这是指用 ubuntu 15.10 版本镜像为基础来启动容器。

/bin/bash:放在镜像名后的是命令,这里希望有个交互式 Shell,因此用的是/bin/bash。

如果要使用版本为 14.04 的 ubuntu 系统镜像来运行容器,命令如下:docker run -t -i ubuntu:14.04 /bin/bash。

如果不指定一个镜像的版本标签,例如只使用 ubuntu,docker 将默认使用 ubuntu:latest 镜像。

2. 获取一个新的镜像

当在本地主机上使用一个不存在的镜像时,Docker 就会自动下载这个镜像。如果想预先下载这个镜像,可以使用 docker pull 命令来下载它。如:

```
[root@localhost heathcliff] #docker pull ubuntu:13.10
13.10: pulling from library/ubuntu
Image docker.io/library/ubuntu:13.10 uses outdated schemal manifest format. Please upgrade to a schema2 image for better future compatibility. More information at https://docs.docker.com/registry/spec/deprecated-schema-v1/
a3ed95caeb02: Pull complete
0d8710fc57fd: Pull complete
5037c5cd623d: Pull complete
83b53423b49f: Pull complete
e9e8bd3b94ab: Pull complete
7db00e6b6e5e: Pull complete
Digest: sha256:403105e61e2d540187da20d837b6a6e92efc3eb4337da9c04c191fb5e28c44dc
Status: Downloaded newer image for ubuntu:13.10
docker.io/library/ubuntu:13.10
```

下载完成后,可以直接使用这个镜像来运行容器。

3. 查找镜像

可以用 Docker Hub 网站来搜索镜像,Docker Hub 网址为"https://hub.docker.com/",也可以使用 docker search 命令来搜索镜像。比如需要一个 httpd 的镜像来作为 Web 服务,可以通过 docker search 命令搜索 httpd 来寻找适合的镜像。

```
docker search httpd
```

NAME:镜像仓库源的名称。

DESCRIPTION:镜像的描述。

STARS:类似 Github 里面的 star,表示点赞、喜欢的意思。

OFFICIAL:是否为 docker 官方发布。

AUTOMATED:自动构建。

```
[root@localhost heathcliff]# docker search httpd
NAME                              DESCRIPTION                                    STARS    OFFICIAL   AUTO
Httpd                             The Apache HTTP Server Project                 3185     [OK]
centos/httpd-24-centos7           Platform for running Apache httpd 2.4--        36
centos/httpd                                                                     32                  [OK]
arm32v7/httpd                     The Apache HTTP Server Project                 9
polinux/httpd-php                 Apache with PHP in Docker (Supervisor, Cent0-  4                   [OK]
saliml983hoop/httpd24             Dockerfile running apache config               2                   [OK]
clearlinux/httpd                  httpd HyperText Transfer Protocol (HTTP) ser...1
publici/httpd                     httpd:latest                                   1                   [OK]
solsson/httpd-openidc             mod_auth_openidc on official httpd image, ve...1                   [OK]
jonathanheilmann/httpd-           httpd:alpine with enabled mod rewrite          1                   [OK]
lead4good/httpd-fpm               httpd server which connects via fcgi proxy h...1                   [OK]
dariko/httpd-rproxy-ldap          Apache httpd reverse proxy with LDAP authent...1                   [OK]
dockerpinata/httpd                                                               0
e2eteam/httpd                                                                    0
interlutions/httpd                httpd docker image with debian-based config... 0                   [OK]
appertly/httpd                    Customized Apache HTTPD that uses a PHP-FPM... 0                   [OK]
amd64/httpd                       The Apache HTTP Server Project                 0
manasip/httpd                                                                    0
trollin/httpd                                                                    0
hypoport/httpd-cgi                httpd-cgi                                      0                   [OK]
manageiq/httpd configmap generator Httpd Configmap Generator                     0                   [OK]
itsziget/httpd24                  Extended HTTPD Docker image based on the f...  0                   [OK]
alvistack/httpd                   Docker Image Packaging for Apache              0                   [OK]
manageiq/httpd                    Container with httpd, built on Centos for Ma...0                   [OK]
ppc64le/httpd                     The Apache HTTP Server Project                 0
```

4. 拖取镜像

若决定使用上图中的 httpd 官方版本的镜像,使用命令 docker pull 来下载镜像。

```
docker pull httpd
```

```
[root@localhost heathcliff]# docker pull httpd
using default tag:latest
```

```
latest:Pulling from library/httpd
d121f8d1c412:Pull complete
9cd35c2006cf:Pull complete
b6b9dec6e0f8:Pull complete
fc3f9b55fcc2:Pull complete
802357647f64:Pull complete
Digest:sha256:5ce7c20e45b407607f30b8f8ba435671c2ff80440d12645527be670eb8ce1961
Status:DownLoaded newer image for httpd:latest
docker.io/library/httpd:latest
```

下载完成后,就可以使用这个镜像了。

```
docker run httpd
```

5. 删除镜像

镜像删除使用 docker rmi 命令,比如删除 hello-world 镜像。

```
docker rmi -f hello-world
```

-f 表示强制删除。

```
[root@localhost heathcliff]# docker rmi -f hello-world
Untagged:hello-world:latest
Untagged:hello-world@sha256:4cf9c47f86df71d48364001ede3a4fcd85ae80ce02ebad74156906caff5378bc
Deleted:sha256:bf756fb1ae65adf866bd8c456593cd24beb6 a0a061dedf42b26a993176745f6b
```

6. 创建更改镜像

当从 docker 镜像仓库中下载的镜像不能满足需求时,可以通过以下两种方式对镜像进行更改。

(1)从已经创建的容器中更新镜像,并且提交这个镜像。

(2)使用 Dockerfile 指令来创建一个新的镜像。

更新镜像之前,需要使用镜像来创建一个容器。

```
[root@localhost heathcliff]# docker run -t -i ubuntu:15.10  /bin/bash
root@e02cb5d524d9:/#
```

在运行的容器内使用 apt-get update 命令进行更新。

在完成操作之后,输入 exit 命令来退出这个容器。

此时 ID 为"e02cb5d524d9"的容器为按需求更改的容器,可以通过命令 docker commit 来提交容器副本。

```
docker commit -m="has update" -a="heathcliff" e02cb5d524d9 heathcliff/ubuntu:v2
```

各个参数说明:

-m：提交的描述信息。

-a：指定镜像作者。

e02cb5d524d9：容器 ID。

heathcliff/ubuntu：v2：指定要创建的目标镜像名。

```
[root@localhost heathcliff] # docker commit -m="has update" -a="heathcliff" e02cb5d524d9 heathcliff/ubuntu:v2
sha256:9586cb9134ead096244c58d6dc08fa35e0403d143af57a90e0ce4 cb8d42b4242
```

可以使用 docker images 命令来查看我们的新镜像"heathcliff/ubuntu v2"。

```
[root@localhost heathcliff] # docker images
REPOSITORY           TAG        IMAGE ID        CREATED          SIZE
heathcliff/ubuntu    v2         9586cb9134ea    17 seconds ago   137MB
ubuntu               latest     9140108b62dc    3 days ago       72.9MB
httpd                latest     417af7dc28bc    13 days ago      138MB
ubuntu               15.10      9b9cb95443b5    4 years ago      137MB
training/webapp      latest     6fae60ef3446    5 years ago      349MB
ubuntu               13.10      7f020f7bf345    6 years ago      185MB
```

使用新镜像 heathcliff/ubuntu 来启动一个容器：

```
[root@localhost heathcliff] # docker run -t -i heathcliff/ubuntu:v2  /bin/bash
root@75b93a124778:/#
```

三、Docker Dockerfile 构建镜像

Dockerfile 是一个用来构建镜像的文本文件，文本内容包含了一条条构建镜像所需的指令和说明。可以通过运行 Dockerfile 文件来定制一个镜像。

1. 定制一个 nginx 镜像

构建好的镜像内会有一个/usr/share/nginx/html/index.html 文件。

在一个空目录下，新建一个名为"Dockerfile"的文件，并在文件内添加以下内容。

```
FROM nginx
RUN echo '这是一个本地构建的 nginx 镜像' >/usr/share/nginx/html/index.html
```

如：

```
[root@localhost heathcliff] # mkdir Dockerfile
[root@localhost heathcliff] # cd Dockerfile/
[root@localhost Dockerfile] # vi Dockerfile
[root@localhost Dockerfile] # cat Dockerfile
FROM nginx
RUN echo '这是一个本地构建的 nginx 镜像'>/usr/share/nginx/html/index.html
```

2. FROM 和 RUN 指令的作用

FROM：定制的镜像都是基于 FROM 的镜像，这里 nginx 就是定制需要的基础镜像，后续的操作都是基于 nginx。

RUN：用于执行后面跟着的命令行命令，有以下两种格式。

shell 格式：

```
RUN <命令行命令>
# <命令行命令>等同于在终端操作的 shell 命令。
```

exec 格式：

```
RUN ["可执行文件", "参数 1", "参数 2"]
# 例如：RUN ["./test.php", "dev", "offline"] 等价于 RUN ./test.php dev offline
```

注意：Dockerfile 的指令每执行一次都会在 docker 上新建一层。所以，过多无意义的层会造成镜像膨胀过大。例如：

```
FROM centos
RUN yum install wget
RUN wget -O redis.tar.gz "http://download.redis.io/releases/redis-5.0.3.tar.gz"
RUN tar -xvf redis.tar.gz
以上执行会创建 3 层镜像。可简化为以下格式：
FROM centos
RUN yum install wget\
&& wget -O redis.tar.gz "http://download.redis.io/releases/redis-5.0.3.tar.gz"\
&& tar -xvf redis.tar.gz
```

如上，以 && 符号连接命令，这样执行后，只会创建一层镜像。

在"Dockerfile"文件的存放目录下，执行构建动作。

以下示例，通过目录下的 Dockerfile 构建一个"nginx:test"（镜像名称:镜像标签）。

注：最后的 . 代表本次执行的上下文路径"docker build -t nginx:v3"，当出现以下提示，说明已经构建成功。

```
[root@localhost Dockerfile] # docker build -t nginx:v3
Sending build context to Docker daemon 2.048KB
Step1/2：FROM nginx
latest：Pulling from library/nginx
d121f8d1c412：Already exists
66a200539fd6：Pull complete
e9738820db15：Pull complete
d74ea5811e8a：Pull complete
ffdacbba6928：Pull complete
Digest：sha256:fc66cdef5ca33809823182c9c5d72ea86fd2cef7713cf3363ela0b12a5d77500
```

```
Status：Downloaded newer image for nginx:latest
--->992e3b7be046
step 2/2：RUN echo"这是一个本地构建的 nginx 镜像">/usr/share/nginx/html/index.html
--->Running in 488391c5bf4f
Removing intermediate container 488391c5bf4f
--->1471fe752e01
Successfully built 1471fe752e01
Successfully tagged nginx：v3
[root@localhost Dockerfile]# docker images
REPOSITORY          TAG          IMAGE ID          CREATED          SIZE
nginx               v3           1471fe752e01      59 seconds ago   133MB
```

前面提到指令最后一个．是上下文路径。上下文路径，是指 docker 在构建镜像时，有时候想要使用到本机的文件（比如复制），docker build 命令得知这个路径后，会将路径下的所有内容打包。

解析：由于 docker 的运行模式是 C/S，那么本机是 C，docker 引擎是 S。实际的构建过程是在 docker 引擎下完成的，所以这个时候无法用到本机的文件。这就需要把本机的指定目录下的文件一起打包提供给 docker 引擎使用。

如果未说明最后一个参数，那么默认上下文路径就是 Dockerfile 所在的位置。

注意：上下文路径下不要放无用的文件，因为会一起打包发送给 docker 引擎，如果文件过多，会造成过程缓慢。

3. 其他指令的使用

1）COPY

复制指令，从上下文目录中复制文件或者目录到容器里的指定路径。

格式：

```
COPY [--chown=<user>:<group>] <源路径1>...   <目标路径>
COPY [--chown=<user>:<group>] ["<源路径1>",...   "<目标路径>"]
```

[--chown=<user>:<group>]：可选参数，用户改变复制到容器内文件的拥有者和属组。

<源路径>：源文件或者源目录，这里可以是通配符表达式，其通配符规则要满足 Go 的 filepath.Match 规则。例如：

```
COPY  hom*  /mydir/
COPY  hom?.txt  /mydir/
```

<目标路径>：容器内的指定路径，该路径不用事先建好，路径不存在的话会自动创建。

2）ADD

ADD 指令和 COPY 的使用格式一致，功能也类似，不同之处如下。

ADD 的优点：在执行<源文件>为 tar 压缩文件的话，压缩格式为 gzip；bzip2 以及 xz 的情况下，会自动复制并解压到<目标路径>。

ADD 的缺点：在不解压的前提下，无法复制 tar 压缩文件。会令镜像构建缓存失效，从而可能会令镜像构建变得比较缓慢。具体是否使用，可以根据是否需要自动解压来决定。

3）CMD

CMD 类似于 RUN 指令，用于运行程序，但二者运行的时间点不同，CMD 在 docker run 时运行，RUN 在 docker build 时运行。

作用：为启动的容器指定默认要运行的程序，程序运行结束，容器也就结束。CMD 指令指定的程序可被 docker run 命令行参数中指定要运行的程序所覆盖。

注意：Dockerfile 中如果存在多个 CMD 指令，仅最后一个生效。

格式：

```
CMD <shell 命令>
CMD ["<可执行文件或命令>","<param1>","<param2>",...]
CMD ["<param1>","<param2>",...]   #该写法是为 ENTRYPOINT 指令指定的程序提供默认参数
```

推荐使用第二种格式，执行过程比较明确。第一种格式实际上在运行的过程中也会自动转换成第二种格式运行，并且默认可执行文件是 sh。

4）ENTRYPOINT

ENTRYPOINT 类似于 CMD 指令，但其不会被 docker run 的命令行参数指定的指令所覆盖，而且这些命令行参数会被当作参数送给 ENTRYPOINT 指令指定的程序。

但是，如果运行 docker run 时使用了 --entrypoint 选项，此选项的参数可当作要运行的程序覆盖 ENTRYPOINT 指令指定的程序。

优点：在执行 docker run 的时候可以指定 ENTRYPOINT 运行所需的参数。

注意：Dockerfile 中如果存在多个 ENTRYPOINT 指令，仅最后一个生效。

格式：

```
ENTRYPOINT ["<executeable>","<param1>","<param2>",...]
```

可以搭配 CMD 命令使用：一般是变参才会使用 CMD，这里的 CMD 等于是在给 ENTRYPOINT 传参，以下示例会提到。

示例：假设已通过 Dockerfile 构建了 nginx:test 镜像。

```
FROM nginx
ENTRYPOINT ["nginx", "-c"] #定参
CMD ["/etc/nginx/nginx.conf"] #变参
```

其中，不传参运行：docker run nginx:test。

容器内会默认运行以下命令，启动主进程，nginx -c /etc/nginx/nginx.conf。

传参运行：docker run nginx:test -c /etc/nginx/new.conf。

容器内会默认运行以下命令，启动主进程（/etc/nginx/new.conf：假设容器内已有此文件）：nginx -c /etc/nginx/new.conf。

5）ENV

ENV 设置环境变量，定义了环境变量，那么在后续的指令中就可以使用这个环境变量。

格式：

```
ENV <key> <value>
ENV <key1>=<value1> <key2>=<value2>...
```

以下示例设置 NODE_VERSION=7.2.0，在后续指令中可以通过 $NODE_VERSION 引用。

```
ENV NODE_VERSION 7.2.0
RUN curl -SLO "https://nodejs.org/dist/v$NODE_VERSION/node-v$NODE_VERSION-linux-x64.tar.xz" \&& curl -SLO "https://nodejs.org/dist/v$NODE_VERSION/SHASUMS256.txt.asc"
```

6）ARG

ARG 构建参数与 ENV 作用一致，不过作用域不一样。ARG 设置的环境变量仅对 Dockerfile 内有效，也就是说只有在 docker build 的过程中有效，构建好的镜像内不存在此环境变量。

构建命令 docker build 中可以用--build-arg <参数名>=<值> 来覆盖。

格式：ARG <参数名>[=<默认值>]。

7）VOLUME

VOLUME 定义匿名数据卷。在启动容器时忘记挂载数据卷，会自动挂载到匿名卷。

作用：①避免重要的数据因容器重启而丢失（数据丢失是非常致命的）；②避免容器不断变大。

格式：

```
VOLUME ["<路径1>", "<路径2>"...]
VOLUME <路径>
```

在启动容器 docker run 的时候，可以通过-v 参数修改挂载点。

8）EXPOSE

EXPOSE 仅仅只是声明端口。

作用：①帮助镜像使用者理解这个镜像服务的守护端口，以方便配置映射；②在运行时使用随机端口映射，也就是 docker run -P 时，会自动随机映射 EXPOSE 的端口。

格式：EXPOSE <端口1> [<端口2>...]。

9）WORKDIR

WORKDIR 指定工作目录。用 WORKDIR 指定的工作目录，会在构建镜像的每一层中都存在（WORKDIR 指定的工作目录，必须是提前创建好的）。

docker build 构建镜像过程中的每个 RUN 命令都是新建的一层，只有通过 WORKDIR 创建的目录才会一直存在。

格式：WORKDIR <工作目录路径>。

10）USER

USER 用于指定执行后续命令的用户和用户组，这里只是切换后续命令执行的用户（用户和用户组必须提前已经存在）。

格式：USER <用户名>[:<用户组>]。

11) HEALTHCHECK

HEALTHCHECK 用于指定某个程序或者指令来监控 docker 容器服务的运行状态。

格式：

```
HEALTHCHECK [选项] CMD <命令>:设置检查容器健康状况的命令
HEALTHCHECK NONE:如果基础镜像有健康检查指令,使用这行可以屏蔽其健康检查指令
HEALTHCHECK [选项] CMD <命令> :这里 CMD 后面跟随的命令使用,可以参考 CMD 的用法
```

12) ONBUILD

ONBUILD 指令实际上相当于创建一个模板镜像,后续可以根据该模板镜像创建特定的子镜像。在利用该 Dockerfile 构建父镜像（如 A 镜像）时,该指令不会执行。但是当编写一个新的 Dockerfile 文件来基于 A 镜像构建子镜像（如 B 镜像）时,这时构造 A 镜像时用的 Dockerfile 文件中的 ONBUILD 指令就会执行。但是由子镜像再创建孙镜像时,ONBUILD 指令不再执行。

格式：ONBUILD <其他指令>。

四、Docker 容器连接

容器中可以运行一些网络应用,要让外部也可以访问这些应用,可以通过 -P 或 -p 参数来指定端口映射。下面来实现通过端口连接到一个 docker 容器。

1. 网络端口映射

先创建一个 python 应用的容器：

```
[root@localhost heathcliff]# docker run -d -P training/webapp python app.py
4a5680221844e6f2674b8ed14c6c1429f5b9e8a7803b16356d3ed581 d5e754d0
```

另外,可以指定容器绑定的网络地址,比如绑定 127.0.0.1。

使用 -P 参数创建一个容器,用 docker ps 可以看到容器端口 5000 绑定主机端口 32768。

```
[root@localhost heathcliff]# docker ps
CONTAINER ID   IMAGE            COMMAND           CREATED           STATUS              PORTS                     NAMES
4a5680221844   training/webapp  "python app.py"   About a minute ago Up About a minute   0.0.0.0:32768->5000/tcp   funny_austin
```

另外,也可以使用 -p 标识来指定容器端口绑定到主机端口：

```
docker run -d -p 5000:5000 training/webapp python app.py
```

两种方式的区别是：-P 是容器内部端口随机映射到主机的高端口,-p 是容器内部端口绑定到指定的主机端口。

```
[root@localhost heathcliff]# docker run -d -p 5000:5000 training/webapp python app.py
2b3a99c9493dde93e50745a2fe2 a595edd635c277dac65a83af1 b0781f31dl fd
[root@localhost heathcliff]# docker ps
CONTAINER ID   IMAGE            COMMAND           CREATED          STATUS         PORTS                     NAMES
2b3a99c9493d   training/webapp  "python app.py"   22 seconds ago   Up 21 seconds  0.0.0.0:5000->5000/tcp    thirsty_cannon
4a5680221844   training/webapp  "python app.py"   13 hours ago     Up 13 hours    0.0.0.0:32768->5000/tcp   funny_austin
```

另外，可以指定容器绑定的网络地址，比如绑定 127.0.0.1：

```
[root@localhost heathcliff]# docker run -d -p 127.0.0.1:5001:5000 training/webapp python app.py
d0b3f617e6e8d3af3601381945e4722a43c38bae98ea41445a1e59ed31340623
[root@localhost heathcliff]# docker ps
CONTAINER ID    IMAGE            COMMAND            CREATED             STATUS            PORTS                       NAMES
d0b3f617e6e8    training/webapp  "python app.py"    3 seconds ago       Up 3 seconds      127.0.0.1:5001-5000/tcp     zealous_sanderson
2b3a99c9493d    training/webapp  "python app.py"    About a minute ago  Up About a minute 0.0.0.0:5000-5000/tcp       thirsty_cannon
4a5680221844    training/webapp  "python app.py"    13 hours ago        Up 13 hours       0.0.0.0:32768-5000/tcp      funny_austin
```

这样就可以通过访问 127.0.0.1:5001 来访问容器的 5000 端口。

上面的例子中，默认都是绑定 tcp 端口，如果要绑定 UDP 端口，可以在端口后面加上 /udp：

```
[root@localhost heathcliff]# docker run -d -p 127.0.0.1:5000:5000/udp training/webapp python app.py
b05305bb3136dc802a75879fc16c2c6f591153d075cc050bd8f1ba27d981df2b
[root@localhost heathcliff]# docker ps
CONTAINER ID    IMAGE            COMMAND            CREATED          STATUS         PORTS                                   NAMES
b05305bb3136    training/webapp  "python app.py"    4 seconds ago    Up 3 seconds   5000/tcp,127.0.0.1:5000-5000/udp        nostalgic_wu
d0b3f617e6e8    training/webapp  "python app.py"    3 minutes ago    Up 3 minutes   127.0.0.1:5001-5000/tcp                 zealous_sanderson
2b3a99c9493d    training/webapp  "python app.py"    5 minutes ago    Up 5 minutes   0.0.0.0:5000-5000/tcp                   thirsty_cannon
4a5680221844    training/webapp  "python app.py"    13 hours ago     Up 13 hours    0.0.0.0:32768-5000/tcp                  funny_austin
```

docker port 命令可以快捷地查看端口的绑定情况：

```
[root@localhost heathcliff]# docker port nostalgic_wu
5000/udp->127.0.0.1:5000
```

2. Docker 容器互联

端口映射并不是唯一把 docker 连接到另一个容器的方法。

docker 有一个连接系统允许将多个容器连接在一起，共享连接信息。

docker 连接会创建一个父子关系，其中父容器可以看到子容器的信息。

当创建一个容器的时候，docker 会自动对它进行命名。另外，也可以使用 --name 标识来命名容器，例如：

```
[root@localhost heathcliff]# docker run -d -P --name runoob training/webapp python app.py
e0bcf2eb8bf165cdf0823a98c37313b8752df1 beb8c8c8d7e6631f76bed41587
[root@localhost heathcliff]# docker ps -l
CONTAINER ID    IMAGE            COMMAND            CREATED          STATUS         PORTS                      NAMES
e0bcf2eb8bf1    training/webapp  "python app.py"    9 seconds ago    Up 8 seconds   0.0.0.0:32771->5000/tcp    runoob
```

下面先创建一个新的 Docker 网络：

```
docker network create -d bridge test-net
```

参数说明：

-d：参数指定 Docker 网络类型，有 bridge、overlay。

其中 overlay 网络类型用于 Swarm mode，在本小节中可以忽略它。

```
[root@localhost heathcliff]# docker network create -d bridge test-net
5694e58e9dff704cbldf61bb59acbb9bfc4db568c8ee930677816a5fba70b223
[root@localhost heathcliff]# docker network ls
NETWORK ID          NAME                DRIVER              SCOPE
b6b4413e4087        bridge              bridge              local
45921fd5c2c8        composetest_default bridge              local
836823a2a3f7        host                host                local
77ca2be5b628        none                null                local
5694e58e9dff        test-net            bridge              local
```

运行一个容器并连接到新建的 test-net 网络：

```
[root@localhost heathcliff]# docker run -itd --name test1 --network test-net ubuntu /bin/bash
4475ed44ed8b893231b14501accf5367530dde39cc3e873df69fee5c9421a134
```

打开新的终端，再运行一个容器并加入到 test-net 网络：

```
[root@localhost heathcliff]# docker run -itd --name test2 --network test-net ubuntu /bin/bash
e6bb861 8b17039660cbb89c92a6d050b52daedbff91615a507dd0b1043af9140
[root@localhost heathcliff]# docker ps
CONTAINER ID   IMAGE    COMMAND      CREATED           STATUS              PORTS    NAMES
e6bb8618b170   ubuntu   "/bin/bash"  18 seconds ago    Up 17 seconds                test2
4475ed44ed8b   ubuntu   "/bin/bash"  About a minute ago Up About a minute           test1
```

下面通过 ping 来证明 test1 容器和 test2 容器建立了互联关系。

如果 test1、test2 容器内无 ping 命令，则在容器内执行以下命令安装 ping（可以在一个容器里安装好，提交容器到镜像，再以新的镜像重新运行以上两个容器）。

```
apt-get update
apt install iputils-ping
```

在容器 test1 中，用 ping 命令验证与容器 test2 的连接：

```
[root@localhost heathcliff]# docker exec -it test1/bin/bash
root@4475ed44ed8b:/# ping test2
bash:ping:command not found
root@4475ed44ed8b:/# apt-get update
Get:1 http://archive.ubuntu.com/ubuntu focal InRelease [265 KB]
Get:2 http://security.ubuntu.com/ubuntu focal-security InRelease [107 KB]
Get:3 http://security.ubuntu.com/ubuntu focal-security/main amd64 Packages [406 KB]Get:4 http://archive.ubuntu.com/ubuntu focal-updates InRelease [111 KB]
Get:5 http://archive.ubuntu.com/ubuntu focal-backports InRelease [98.3 KB]Get:6 http://archive.ubuntu.com/ubuntu focal/main amd64 Packages [1275 KB]
```

Get:7 http://security.ubuntu.com/ubuntu focal-security/universe amd64 Packages [626 KB]Get:8 http://security.ubuntu.com/ubuntu focal-security/restricted amd64 Packages [75.9KB]
Get:9 http://security.ubuntu.com/ubuntu focal-security/multiverse amd64 Packages [1169 B]Get:10 http://archive.ubuntu.com/ubuntu focal/multiverse amd64 Packages [177 KB]
Get:11 http://archive.ubuntu.com/ubuntu focal/universe amd64 Packages [11.3 NB]Get:12 http://archive.ubuntu.com/ubuntu focal/restricted amd64 Packages [33.4 KB]
Get:13 http://archive.ubuntu.com/ubuntu focal-updates/universe amd64 Packages [832 KB]
Get:14 http://archive.ubuntu.com/ubuntu focal-updates/restricted amd64 Packages [88.7 KB]
Get:15 http://archive.ubuntu.com/ubuntu focal-updates/multiverse amd64 Packages [21.6 KB]Get:16 http://archive.ubuntu.com/ubuntu focal-updates/main amd64 Packages [745 KB]
Get:17 http://archive.ubuntu.com/ubuntu focal-backports/universe amd64 Packages [4277 B]
Fetched 16.2 MB in 1min 14s(220 KB/s)
Reading package lists... Done
root@4475ed44ed8b:/# apt install iputils-ping
Reading package lists... Done
Building dependency tree
Reading state information... Done
The following additional packages will be installed:
libcap2 libcap2-bin libpam-cap
The following NEW packages will be installed:
iputils-ping libcap2 libcap2-bin libpam-cap
0 upgraded,4newly installed,0 to remove and 4 not upgraded.
Need to get 90.5 KB of archives.
After this operation,333 KB of additional disk space will be used.
Do you want to continue? [Y/n] y

在容器 test1 中，用 ping 命令验证与容器 test2 的连接：

root@ 4475ed44ed8b:/# ping test2
PING test2(172.19.0.3)56(84) bytes of data.
64 bytes from test2.test-net (172.19.0.3): icmp_seq=1 ttl=64 time=0.091 ms
64 bytes from test2.test-net (172.19.0.3): icmp_seq=2 ttl=64 time=0.050 ms
64 bytes from test2.test-net (172.19.0.3): icmp_seq=3 ttl= 64 time=0.050 ms
64 bytes from test2.test-net (172.19.0.3): icmp_seq=4 ttl=64 time=0.053 ms
64 bytes from test2.test-net (172.19.0.3): icmp_seq=5 ttl=64 time=0.050 ms

同理在 test2 容器也会成功连接到 test1：

root @ e6bb8618b170:/# ping test1
PING test1 (172.19.0.2)56(84) bytes of data.
64 bytes from test1.test-net (172.19.0.2): icmp_seq=1 ttl=64 time=0.075ms
64 bytes from test1.test-net (172.19.0.2): icmp_seq=2 ttl=64 time=0.050 ms
64 bytes from test1.test-net (172.19.0.2): icmp_seq=3 ttl=64 time=0.051 ms
64 bytes from test1.test-net (172.19.0.2): icmp_seq=4 ttl=64 time=0.051 ms

这样,test1 容器和 test2 容器建立了互联关系。如果有多个容器之间需要互相连接,推荐使用 Docker Compose。

可以在宿主机的 /etc/docker/daemon.json 文件中增加以下内容来设置全部容器的 DNS:

```
[root@localhost heathcliff]# vi /etc/docker/daemon.json
{
  "dns" : [
    "114.114.114.114",
    "8.8.8.8" ]
}
```

设置后,启动容器的 DNS 会自动配置为 114.114.114.114 和 8.8.8.8。

配置完,需要重启 docker 才能生效。

查看容器的 DNS 是否生效可以使用以下命令,输出容器的 DNS 信息:

```
[root@localhost heathcliff]# systemctl restart docker
[root@localhost heathcliff]# docker run -it --rm ubuntu cat etc/resolv.conf
nameserver 114.114.114.114
nameserver 8.8.8.8
```

如果只想在指定的容器设置 DNS,手动指定容器的配置,则可以使用以下命令:

```
docker run -it --rm -h host_ubuntu  --dns=114.114.114.114 --dns-search=test.com ubuntu
```

参数说明:

--rm:容器退出时自动清理容器内部的文件系统。

-h HOSTNAME 或者 --hostname=HOSTNAME:设定容器的主机名,它会被写到容器内的 /etc/hostname 和 /etc/hosts。

--dns=IP_ADDRESS:添加 DNS 服务器到容器的 /etc/resolv.conf 中,让容器用这个服务器来解析所有不在 /etc/hosts 中的主机名。

--dns-search=DOMAIN:设定容器的搜索域,当设定搜索域为"example.com"时,在搜索一个名为 host 的主机时,DNS 不仅搜索 host,还会搜索 host.example.com。

```
[root@localhost heathcliff]# docker run -it --rm -h host_ubuntu --dns=114.114.114.114 --dns-search=test.com ubuntu
root@host_ubuntu:/# cat /etc/hostname
host_ubuntu
root@host_ubuntu:/# cat /etc/hosts
127.0.0.1 localhost
::1       localhost ip6-localhost ip6-loopback
fe00::0   ip6-localnet
ff00::0   ip6-mcastprefix
ff02::1   ip6-allnodes
```

```
ff02::2         ip6-allrouters
172.17.0.2      host_ubuntu
root@host_ubuntu:/# cat etc/resolv.conf
search test.com
nameserver 114.114.114.114
```

如果在容器启动时没有指定--dns 和--dns-search，Docker 会默认用宿主主机上的/etc/resolv.conf 来配置容器的 DNS。

4.3 Docker 应用开发实验

4.3.1 实验目的

（1）熟悉 Docker 安装系统。
（2）熟悉 Docker 安装数据库。
（3）熟悉 Docker 安装 Python。

4.3.2 实验环境

（1）硬件：PC 机。
（2）软件：CentOS 7 及以上系统、Windows 7 及以上系统。
（3）正常互联网网络连接。

4.3.3 实验内容

（1）Docker 安装系统。
（2）Docker 安装数据库。
（3）Docker 安装 Python。

4.3.4 实验步骤

一、Docker 容器安装 Linux 系统

使用 Docker，只需要一个命令就能快速获取一个 Linux 发行版镜像，这是以往各种虚拟化技术都难以实现的。这些镜像一般都很精简，但可以支持完整 Linux 系统的大部分功能。Ubuntu 是基于 Debian 的 Linux 操作系统，下面介绍使用 Docker 容器安装 Ubuntu 系统。

1. 查看可用的 Ubuntu 版本

访问 Ubuntu 镜像库地址：https://hub.docker.com/_/ubuntu?tab=tags&page=1（图 4-11）。

图 4-11 Docker Hub 上的 ubuntu 镜像

可以通过 Sort by 查看其他版本的 ubuntu(图 4-12)。默认是最新版本 ubuntu:latest。

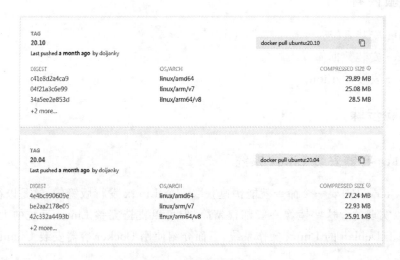

图 4-12 Docker Hub 上不同版本的 ubuntu 镜像

也可以在下拉列表中找到其他想要的版本。

2. 拉取最新版的 Ubuntu 镜像

通过 docker pull ubuntu 或 docker pull ubuntu:latest 命令拉取最新版的 Ubuntu 镜像。

```
docker pull ubuntu
docker pull ubuntu:latest
```

如：

```
[root@localhost heathcliff]# docker pull ubuntu
Using default tag: latest
latest: Pulling from library/ubuntu
Digest: sha256:bc2f7250f69267c9c6b66d7b6a81a54d3878bb85f1ebb5f951c896d13e6ba537
Status: Image is up to date for ubuntu: latest
docker.io/library/ubuntu: latest
```

3. 查看本地镜像

查看本地镜像指令：

[root@localhost heathcliff] # docker images				
REPOSITORY	TAG	IMAGE ID	CREATED	SIZE
heathcliff/ubuntu	v2	9586cb9134ea	20 hours ago	137MB
ubuntu	latest	9140108b62dc	4 days ago	72.9MB
httpd	latest	417af7dc28bc	2 weeks ago	138MB
ubuntu	15.10	9b9cb95443b5	4 years ago	137MB
training/webapp	latest	6fae60ef3446	5 years ago	349MB
ubuntu	13.10	7f020f7bf345	6 years ago	185MB

在上图中可以看到已经安装了最新版本的 ubuntu。

4. 运行容器

运行容器，并且可以通过 exec 命令进入 ubuntu 容器：

```
[root@localhost heathcliff]# docker run -itd --name ubuntu-test ubuntu
479ba9aa07a671ef6a28139eladc48dc19cf844a04a0ff1b8cc31f3957f6dlf2
[root@localhost heathcliff]# docker exec -it ubuntu-test /bin/bash
root@479ba9aa07a6:/#
```

5. 安装成功

最后可以通过 docker ps 命令查看容器的运行信息：

[root@localhost heathcliff] # docker ps						
CONTAINER ID	IMAGE	COMMAND	CREATED	STATUS	PORTS	NAMES
479ba9aa07a6	ubuntu	"/bin/bash"	2 minutes ago	Up 2 minutes		ubuntu-test

二、Docker 安装 MySQL

MySQL 是世界上最受欢迎的开源数据库。凭借其可靠性、易用性和性能，MySQL 已成为 Web 应用程序的数据库优先选择。

1. 查看可用的 MySQL 版本

访问 MySQL 镜像库地址：https://hub.docker.com/_/mysql?tab=tags（图 4-13）。

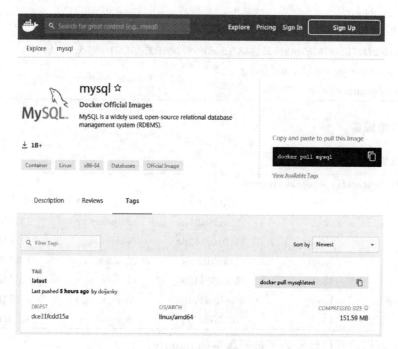

图 4-13　Docker Hub 上的 MySQL 镜像

可以通过 Sort by 查看其他版本的 MySQL，默认是最新版本 mysql:latest（图 4-14）。

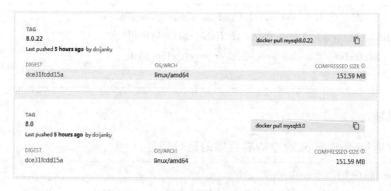

图 4-14　Docker Hub 上不同版本的 MySQL 镜像

也可以在下拉列表中找到其他想要的版本。

此外,还可以用 docker search mysql 命令来查看可用版本:

NAME	DESCRIPTION	STARS	OFFICIAL	AUTO
[root@localhost heathcliff] # docker search mysql				
mysql	MySQL is a widely used,open-source relation…	10003	[OK]	
mariadb	MariaDB is a community-developed fork of MyS…	3662	[OK]	
mysql/mysql-server	Optimized MySQL Server Docker images. Create…	731		[OK]
Percona	Percona Server is a fork of the MySQL relati…	512	[OK]	
centos/mysql-57-centos7	MySQL 5.7 SQL database server	83		
mysql/mysql-cluster	Experimental MySQL Cluster Docker images. Cr…	76		
centurylink/mysql	Image containing mysql. optimized to be link…	61		[OK]
bitnami/mysql	Bitnami MySQL Docker Image	45		[OK]
deitch/mysql-backup	REPLACED! Please use http://hub.docker.com/r…	41		[OK]
tutum/mysql	Base docker image to run a MySQL database se…	35		
prom/mysqld-exporter		31		[OK]
schickling/mysql-backup-s3	Backup MySQL to S3 (supports periodic backup…	30		[OK]
databack/mysql-backup	Back up mysql databases to... anywhere!	30		
linuxserver/mysql	A Mysql container,brought to you by Linuxse…	26		
centos/mysql-56-centos7	MySQL 5.6 SQL database server	20		
circleci/mysql	MySQL is a widely used,open-source relation…	19		
mysql/mysql-router	MySQL Router provides transparent routing be…	16		
arey/mysql-client	Run a MySQL client from a docker container	15		[OK]
fradelg/mysql-cron-backup	MySQL/MariaDB database backup using cron tas…	8		[OK]
openshift/mysql-55-centos7	DEPRECATED: A Centos7 based MySQL v5.5 image…	6		
devilbox/mysql	Retagged MySQL,MariaDB and PerconaDB offici…	3		
ansibleplaybookbundle/mysql-apb	An APB which deploys RHSCL MySQL	2		[OK]
jelastic/mysql	An image of the MySQL database server mainta…	1		
widdpim/mysql-client	Dockerized MySQL client(5.7) including curl…	1		[OK]
monascaf/mysql-init	A minimal decoupled init container for mysql	0		

2. 拉取 MySQL 镜像

这里拉取官方的最新版本的镜像:

```
[root@localhost heathcliff] # docker pull mysql:latest
latest:Pulling from library/mysql
d121f8d1c412:Already exists
f3cebc0b4691 :Pull complete
1862755a0b37: Pull complete
489b44f3dbb4:Pull complete
690874f836db:Pull complete
baa8be383ffb: Pull complete
```

```
55356608b4ac：Pull complete
dd35ceccb6eb：Pull complete
429b35712b19：Pull complete
162d8291095c：Pull complete
5e500ef7181b：Pull complete
af7528e958b6：Pull complete
Digest：sha256：e1bfe11693ed2052cb3b4e5fa356c65381129e87e38551c6cd6ec532ebe0e808
Status：Downloaded newer image for mysql：latest
docker.io/library/mysql：latest
```

3. 查看本地镜像

使用以下命令来查看是否已安装了 mysql(本例中最后一个是 mysql 镜像)：

```
[root@localhost heathcliff] # docker images
REPOSITORY           TAG        IMAGE ID         CREATE          SIZE
heathcliff/ubuntu    v2         9586cb9134ea     21 hours ago    137MB
ubuntu               latest     9140108b62dc     4 days ago      72.9MB
httpd                latest     417af7dc28bc     2 weeks ago     138MB
mysql                latest     e1d7dc9731da     2 weeks ago     544NB
```

在上图中可以看到已经安装了最新版本(latest)的 mysql 镜像。

4. 运行容器

安装完成后，可以使用以下命令来运行 mysql 容器：

```
docker run -itd --name mysql-test -p 3306：3306 -e MYSQL_ROOT_PASSWORD=123456 mysql
```

参数说明：

-p 3306：3306：映射容器服务的 3306 端口到宿主机的 3306 端口，外部主机可以直接通过宿主机 ip：3306 访问到 MySQL 的服务。

MYSQL_ROOT_PASSWORD=123456：设置 MySQL 服务 root 用户的密码。

```
[root@localhost heathcliff] # docker run -itd --name mysql-test -p 3306：3306 -e MYSL_ROOT_PASS-
MORD=123456 mysql
f1806cd10f5aba5a6e313323d5ec7de617b1f55937a5c47c0179ea0fd02137ee
```

5. 安装与运行

通过 docker ps 命令查看是否安装成功：

```
[root@localhost heathcliff] # docker ps
CONTAINER ID   IMAGE    COMMAND              CREATED         STATUS         PORTS                          NAMES
f1806cd10f5a   mysql    "docker-entrypoint.s…"  55 seconds ago  Up 55 seconds  0.0.0.0：3306→3306/tcp,33060/tcp  mysql-test
479ba9aa07a6   ubuntu   "/bin/bash"          2 hours ago     Up 2 hours                                    ubuntu-test
```

本机可以通过 root 和密码 123456 访问 MySQL 服务：

```
[root@localhost heathcliff]# docker exec -it mysql-test bash
root@f1806cd10f5a:/# mysql -u root -p
Enter password：
welcome to the MySQL monitor. Commands end with ; or \g.
Your MySQL connection id is 8
server version：8.0.21 MySQL Community server-GPL
copyright (c) 2000,2020,oracle and/or its affiliates. All rights reserved.
oracle is a registered trademark of Oracle corporation and/or its
affiliates. other names may be trademarks of their respectiveowners.
Type ' help;' or '\h' for help. Type '\c' to clear the current input statement.
```

三、Docker 安装 Python

访问 python 镜像库地址，查找 Docker Hub 上的 python 镜像："https://hub.docker.com/_/python?tab=tags&page=1&ordering=last_updated"（图 4-15）。

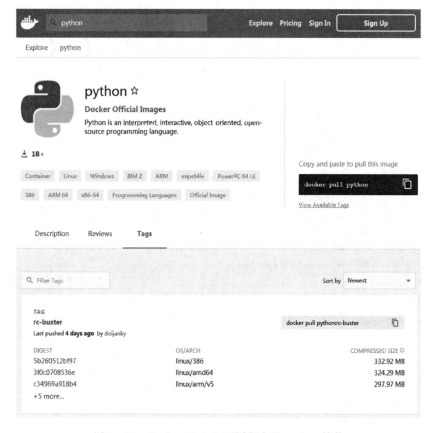

图 4-15　Docker Hub 上不同版本的 python 镜像

可以通过 Sort by 查看其他版本的 python，默认是最新版本 python：lastest。

此外，还可以用 docker search python 命令来查看可用版本：

[root@localhost heathcliff] # docker search python				
NAME	DESCRIPTION	STARS	OFFICIAL	AUTO
python	Python is an interpreted, interactive, objec...	5519	[OK]	
django	Django is a free web application framework,...	999	[OK]	
pypy	PyPy is a fast, compliant alternative implem...	252	[OK]	
nikolaik/python-nodejs	Python with Node.js	53		[OK]
joyzoursky/python-...	Python with chromedriver, for running ...	52		[OK]
arm32v7/python	Python is an interpreted, interactive, objec...	52		
circleci/python	Python is an interpreted, interactive, objec...	39		
centos/python-35-...	Platform for building and running Python 3.5...	38		
centos/python-36-...	Platform for building and running Python 3.6...	30		
Hylang	Hy is a Lisp dialect that translates express...	28	[OK]	
arm64v8/python	Python is an interpreted, interactive, objec...	22		
centos/python-27-centos7	Piatform for building and running Python 2.7...	17		
bitnami/python	Bitnami Python Docker Image	10		[OK]
publicisworldwide/python-conda	Basic Python environments with Conda.	6		[OK]
dockershelf/python	Repository for docker images of Python. Test...	5		[OK]
d3fk/python_in_bottle	Simple python：alpine completed by BottlelReq...	4		[OK]
clearlinux/python	Python programming interpreted language with...	4		
i386/python	Python is an interpreted, interactive, objec...	3		
komand/python-plugin	DEPRECATED：Komand Python SDK	2		[OK]
centos/python-34-centos7	Platform for building and running Python 3.4...	2		
ppc64le/python	Python is an interpreted, interactive, objec...	2		
amd64/python	Python is an interpreted, interactive, objec...	1		
ccitest/python	CirclecI test images for Python	0		[OK]
s390x/python	Python is an interpreted, interactive, objec...	0		
saagie/python	Repo for python jobs	0		

这里拉取官方的镜像，标签为 3.5：

```
[root@localhost heathcliff] # docker pull python：3.5
3.5：Pulling from library/python
57df1alf1ad8：Pull complete
71e126169501：Pull complete
1af28a55c3f3：Pull complete
03f1c9932170：Pull complete
65b3db15f518：Pull complete
850581be87f3：Pull complete
1e37775630ae：Pull complete
7e054ca5fcba：Pull complete
```

92a0fe226896：Pull complete
Digest：sha256：42a37d6b8c00b186bdfb2b620fa8023eb775b3eb3a768fd3c2e421964eee9665
Status：Downloaded newer image for python：3.5
docker.io/library/python：3.5

等待下载完成后，就可以在本地镜像列表里查到 REPOSITORY 为 python、标签为 3.5 的镜像：

```
[root@localhost heathcliff]# docker images python：3.5
REPOSITORY          TAG         IMAGE ID            CREATED             SIZE
python              3.5         3687eb5ea744        3 weeks ago         871MB
```

4.4 Docker 三剑客认识及环境安装实验

4.4.1 实验目的

熟悉 Docker Machine、Docker Compose 和 Swarm 集群管理三剑客与安装。

4.4.2 实验环境

(1) 硬件：PC 机。
(2) 软件：CentOS 7 及以上系统、Windows7 及以上系统。
(3) 正常互联网网络连接。

4.4.3 实验内容

熟悉 Docker Machine、Docker Compose 和 Swarm 集群管理三剑客与安装。

4.4.4 实验步骤

一、Docker Machine

Docker Machine 是一种可以在虚拟主机上安装 Docker 的工具，并可以使用 docker-machine 命令来管理主机。Docker Machine 也可以集中管理所有的 docker 主机，比如快速地给 100 台服务器安装上 docker。

Docker Machine 管理的虚拟主机可以是机上的，也可以是云供应商，如阿里云、腾讯云、AWS 或 DigitalOcean。使用 docker-machine 命令，可以启动、检查、停止和重新启动托管主机，也可以升级 Docker 客户端和守护程序，以及配置 Docker 客户端与使用者的主机进行通信（图 4-16）。

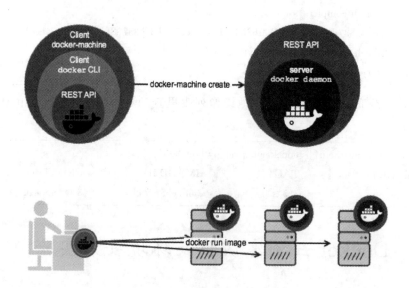

图 4 – 16　Docker Machine 的工作原理

1. 安装

安装 Docker Machine 之前需要先安装 Docker。Docker Machine 可以在多种平台上安装使用,包括 Linux、MacOS 以及 Windows。

(1) Linux 安装命令。

```
[root@localhost heathcliff]# base=https://github.com/docker/machine/releases/download/v0.16.0 &&
>curl-L $base/docker-machine-$(uname -s)-$(uname -m) >/tmp/docker-machine &&
>sudo mv /tmp/docker-machine /usr/local/bin/docker-machine &&
>chmod +x /usr/local/bin/docker-machine
```

%	Total	%	Received	%	Xferd	Average Dload	Speed Upload	Time Total	Time Spent	Time Left	Current Speed
100	651	100	651	0	0	806	0	--:--:--	--:--:--	--:--:--	806
100	26.8M	100	26.8M	0	0	12760	00:36:47	0:36:4	--:--:--		14549

(2) MacOS 安装命令。

```
base=https://github.com/docker/machine/releases/download/v0.16.0 &&
  curl-L $base/docker-machine-$(uname -s)-$(uname -m) >/usr/local/bin/docker-machine &&
  chmod +x /usr/local/bin/docker-machine
```

(3) Windows 安装命令。如果使用 Windows 平台,可以使用 Git BASH,并输入以下命令。

```
base=https://github.com/docker/machine/releases/download/v0.16.0 && \
  mkdir -p "$HOME/bin" && \
  curl -L $base/docker-machine-Windows-x86_64.exe > "$HOME/bin/docker-machine.exe" && \
  chmod +x "$HOME/bin/docker-machine.exe"
```

通过 docker-machine version 指令，查看是否安装成功。

```
[root@localhost heathcliff]# docker-machine version
docker-machine version 0.16.0, build 702c267f
```

2. 使用

通过 virtualbox 来介绍 docker-machine 的使用方法。其他云服务商操作与此基本一致。具体可以参考每家服务商的指导文档。

（1）列出可用的机器。通过 docker-machine ls 指令列出可用的机器。

```
[root@localhost heathcliffl]# docker-machine ls
NAME   ACTIVE   DRIVER   STATE   URL   SWARM   DOCKER   ERRORS
```

（2）创建机器。创建机器前要先安装 virtualbox，首先下载编译 vboxdrv 内核模块所需的构建工具。

```
sudo yum install kernel-devel kernel-headers make patch gcc
```

使用以下 wget 命令将 Oracle Linux repo 文件下载到 /etc/yum.repos.d 目录。

```
[root@localhost heathcliff]# sudo wget https://download.virtualbox.org/virtualbox/rpm/el/virtualbox.repo -P /etc/yum.repos.d
--2020-10-05 10:31:43--https://download.virtualbox.org/virtualbox/rpm/el/virtualbox.repo
正在解析主机 download.virtualbox.org (download.virtualbox.org)... 96.17.188.89
正在连接 download.virtualbox.org (download.virtualbox.org)|96,17,188,89|:443... 已连接
已发出 HTTP 请求,正在等待回应... 200 OK
长度:259[text/plain]
正在保存至:"/etc/yum.repos.d/virtualbox.repo"
100%[===================================>]259        -.-K/s 用时 0s
2020-10-05 10:31:46 (11.0 MB/s)-已保存"etc/yum.repos.d/virtualbox.repo"[259/259]
```

```
[root@localhost heathcliff]# yum install kernel-devel-3.10.0-1127.18.2.el7.x86_64
已加载插件:fastestmirror, langpacks
Loading mirror speeds from cached hostfile
 * base: mirrors.ustc.edu.cn
 * epel: my.mirrors.thegigabit.com
 * extras: mirrors.ustc.edu.cn
 * updates: mirrors.ustc.edu.cn
```

正在解决依赖关系

-->正在检查事务

--->软件包 kernel-devel.x86_64.0.3.10.0-1127.18.2.el7 将被安装

-->解决依赖关系完成

依赖关系解决

Package	架构	版本	源	大小
正在安装：				
kernel-devel	x86_64	3.10.0-1127.18.2.el7	updates	18 M
事务概要				

安装 1 软件包

总下载量：18 M

安装大小：38 M

Is this ok [y/d/N]：y

Downloading packages：

No Presto metadata available for updates

kernel-devel-3.10.0-1127.18.2.el7.x86_64.rpm | 18MB 00：00：01

Running transaction check

Running transaction test

Transaction test succeeded

Running transaction

  正在安装：kernel-devel-3.10.0-1127.18.2.el7.x86_64   1/1

  验证中：kernel-devel-3.10.0-1127.18.2.el7.x86_64   1/1

已安装：

  kernel-devel.x86_64 0：3.10.0-1127.18.2.el7

完毕！

使用 sudo yum install VirtualBox-6.0 命令安装 VirtualBox。安装完毕后出现以下情况则表示成功。

[root@localhost heathcliff] # rcvboxdrv setup

vboxdrv.sh：Stopping VirtualBox services.

vboxdrv.sh：Starting VirtualBox services.

vboxdrv.sh：Building VirtualBox kernel modules.

通过 docker-machine create --driver virtualbox test 创建一台名为"test"的机器。--driver：指定用来创建机器的驱动类型，这里是 virtualbox。

[root@localhost heathcliff] # docker-machine create --driver virtualbox test

Running pre-create checks…

Error with pre-create check："This computer doesn't have VT-X/AND-v enabled. Enabling it in the BIOS is mandatory"

如果没有开启虚拟化引擎,创建会失败,开启虚拟化引擎后重新创建机器(图 4-17)。

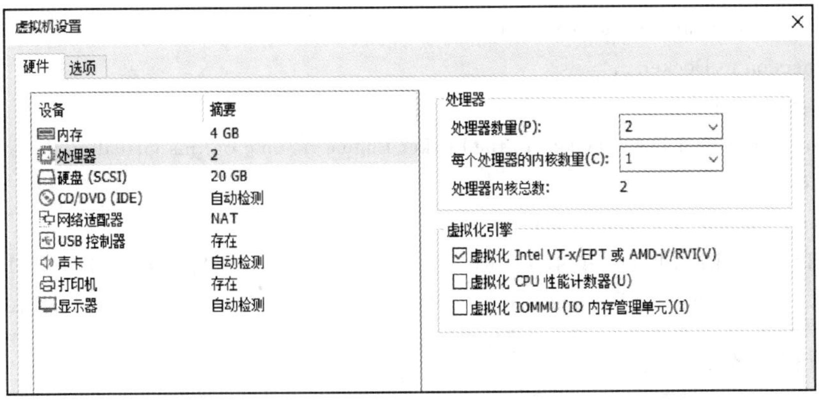

图 4-17 开启虚拟化引擎

```
[root@localhost heathcliff]# docker-machine create --driver virtualbox test
Running pre-create checks...
(test) No default Boot2Docker ISO found locally,downloading the latest release...
(test) Latest release for github.com/boot2docker/boot2docker is v19.03.12
(test) Downloading /root/.docker/machine/cache/boot2docker.iso from https://github.com/boot2docker/
boot2docker/releases/download/v19.03.12/boot2docker.iso...
(test) 0%....10%....20%....30%....40%....50%....60%....70%....80%....90%....100%
creating machine...
(test) Unable to get the latest Boot2Docker ISO release version:Get https://api.github.com/repos/
boot2docker/boot2docker/releases/latest: dial tcp 13.250.168.23:443: connect: connection refused
(test) Copying /root/.docker/machine/cache/boot2docker.iso to /root/.docker/machine/machines/test/
boot2docker.iso...
(test)Creating virtualBox VM...
(test) Creating SSH key...
(test)Starting the VM...
(test) Check network to re-create if needed...
(test) Found a new host-only adapter: "vboxnet0"
(test) Waiting for an IP...
Waiting for machine to be running,this may take a few minutes...
Detecting operating system of created instance...
Waiting for SSH to be available...
Detecting the provisioner...
Provisioning with boot2docker...
```

```
Copying certs to the local machine directory...
Copying certs to the remote machine...
Setting Docker configuration on the remote daemon...
Checking connection to Docker...
Docker is up and running!
To see how to connect your Docker client to the Docker Engine running on this virtual machine, run: docker-machine env test
[root@localhost heathcliff]# docker-machine ls
NAME   ACTIVE   DRIVER       STATE     URL                          SWARM   DOCKER     ERRORS
test   -        virtualbox   Running   tcp://192.168.99.100:2376            v19.03.12
```

(3)查看机器的 ip。

```
[root@localhost heathcliff]# docker-machine ip test
192.168.99.100
```

(4)停止机器。

```
[root@localhost heathcliff]# docker-machine stop test
Stopping "test"...
Machine "test" was stopped.
```

(5)启动机器。

```
[root@localhost heathcliff]# docker-machine start test
Starting "test"...
(test) check network to re-create if needed...
(test) waiting for an IP...
Machine "test" was started.
Waiting for SSH to be available...
Detecting the provisioner...
Started machines may have new IP addresses. You may need to re-run the 'docker-machine env' command.
```

(6)进入机器。

```
[root@localhost heathcliff]# docker-machine ssh test
  ( '>')
  /) TC (\        Core is distributed with ABSOLUTELY NO WARRANTY
 (/-_--_-\)              www.tinycorelinux.net
docker@test:~$
```

二、Docker Compose

Compose 是用于定义和运行多容器 Docker 应用程序的工具。通过 Compose,可以使用 YML 文件来配置应用程序需要的所有服务。然后使用一个命令,就可以从 YML 文件配置中创建并启动所有服务。如果不了解 YML 文件配置,可以先阅读 YAML 入门教程。

使用 Compose 的 3 个步骤:①使用 Dockerfile 定义应用程序的环境;②使用 docker-compose.yml 定义构成应用程序的服务,这样它们可以在隔离环境中一起运行;③执行 docker-compose up 命令来启动并运行整个应用程序。

1. Compose 安装

(1)Linux:在 Linux 上可以从 Github 上下载它的二进制包来使用,最新发行的版本地址:https://github.com/docker/compose/releases。

运行以下命令以下载 Docker Compose 的当前稳定版本:

```
[root@localhost heathcliff]# sudo curl -L "https://github.com/docker/compose/releases/download/1.24.1/docker-compose-$(uname -s)-$(uname -m)" -o /usr/local/bin/docker-compose
```

%	Total	%	Received	%	xferd	Average Dload	Speed Upload	Time Total	Time Spent	Time Left	Current Speed
100	651	100	651	0	0	716	0	--:--:--	--:--:--	--:--:--	716
100	15.4M	100	15.4M	0	0	64553	0	0:04:10	0:04:10	--:--:--	47236

将可执行权限应用于二进制文件:sudo chmod +x /usr/local/bin/docker-compose,或者创建软链接:sudo ln -s /usr/local/bin/docker-compose /usr/bin/docker-compose。

通过 docker-compose --version 测试是否安装成功。

```
[root@localhost heathcliff]# sudo chmod +x /usr/local/bin/docker-compose
[root@localhost heathcliff]# sudo ln -s /usr/local/bin/docker-compose /usr/bin/docker-compose
[root@localhost heathcliff]# docker-compose --version
docker-compose version 1.24.1, build 4667896b
```

注意:对于 alpine,需要以下依赖包:py-pip、python-dev、libffi-dev、openssl-dev、gcc、libc-dev 和 make。

(2)macOS:Mac 的 Docker 桌面版和 Docker Toolbox 已经包括 Compose 及其他 Docker 应用程序,因此 Mac 用户不需要单独安装 Compose。

(3)Windows PC:Windows 的 Docker 桌面版和 Docker Toolbox 已经包括 Compose 及其他 Docker 应用程序,因此 Windows 用户不需要单独安装 Compose。

2. 使用

(1)准备:首先创建一个测试目录。

```
mkdir composetest
cd composetest
```

在测试目录中创建一个名为"app.py"的文件,并复制粘贴以下代码内容:

```
import time
import redis
from flask import Flask

app = Flask(__name__)
cache = redis.Redis(host='redis', port=6379)

def get_hit_count():
    retries = 5
    while True:
        try:
            return cache.incr('hits')
        except redis.exceptions.ConnectionError as exc:
            if retries == 0:
                raise exc
            retries -= 1
            time.sleep(0.5)

@app.route('/')
def hello():
    count = get_hit_count()
    return 'Hello World! I have been seen {} times.\n'.format(count)
```

在此示例中,redis 是应用程序网络上的 redis 容器主机名,该主机使用的端口为 6379。

在 composetest 目录中创建另一个名为 "requirements.txt" 的文件,内容如下:

```
flask
redis
```

(2)创建 Dockerfile 文件。在 composetest 目录中,创建一个名为 "Dockerfile" 的文件,内容如下:

```
FROM python:3.7-alpine
WORKDIR /code
ENV FLASK_APP app.py
ENV FLASK_RUN_HOST 0.0.0.0
RUN apk add --no-cache gcc musl-dev linux-headers
COPY requirements.txt requirements.txt
RUN pip install -r requirements.txt
COPY . .
CMD ["flask", "run"]
```

输入的命令集合如下：

```
[root@localhost heathcliff]# mkdir composetest
[root@localhost heathcliff]# cd composetest
[root@localhost composetest]# rmdir app.py
[root@localhost composetest]# vi app.py
[root@localhost composetest]# vi requirements.txt
[root@localhost composetest]# vi Dockerfile
```

Dockerfile 内容解释：

FROM python:3.7-alpine：从 Python 3.7 映像开始构建镜像。

WORKDIR /code：将工作目录设置为 /code。

设置 flask 命令使用的环境变量。

```
ENV FLASK_APP app.py
ENV FLASK_RUN_HOST 0.0.0.0
```

RUN apk add --no-cache gcc musl-dev linux-headers：安装 gcc，以便诸如 MarkupSafe 和 SQLAlchemy 之类的 Python 包可以编译加速。

复制 requirements.txt 并安装 Python 依赖项。

```
COPY requirements.txt requirements.txt
RUN pip install-r requirements.txt
```

COPY . .：将 . 项目中的当前目录复制到 . 镜像中的工作目录。

CMD ["flask", "run"]：容器提供默认的执行命令为 flask run。

（3）创建 docker-compose.yml。在测试目录中创建一个名为"docker-compose.yml"的文件，然后粘贴以下内容：

```
# yaml 配置
version:'3'
services:
web:
  build:.
  ports:
   -"5000:5000"
redis:
  image:"redis:alpine"
```

该 Compose 文件定义了两个服务：web 和 redis。

web：该 web 服务使用从 Dockerfile 当前目录中构建的镜像。然后，它将容器和主机绑定到暴露的端口 5000。此示例服务使用 Flask Web 服务器的默认端口 5000。

redis：该 redis 服务使用 Docker Hub 的公共 Redis 映像。

(4)使用 Compose 命令构建和运行用户的应用。在测试目录中,执行以下命令来启动应用程序:docker-compose up。若想在后台执行该服务可以加上-d 参数:docker-compose up -d。

```
[root@localhost composetest]# docker-compose up
Creating network "composetest_default" with the default driver
Building web
Step 1/9 : FROM python:3.7-alpine
3.7-alpine: Pulling from library/python
df20fa9351a1:Pull complete
36b3adc4ff6f:Pull complete
4db9de03f499:Pull complete
cd38a04a61f4:Pull complete
9a3838385f13:Pull complete
Digest: sha256:9fbee97d521b846689f4dbf0d5f2770c734d4a09e6d0a0991efc916c58970e99
Status: Downloaded newer image for python:3.7-alpine
--->295b051ee125
Step 2/9 : WORKDIR/code
--->Running in 67dbd14c2c5b
Removing intermediate container 67dbd14c2c5b
---> 62479652bf14
Step 3/9 :ENV FLASK_APP app.py
--->Running in 9b9a9f0867ad
Removing intermediate container 9b9a9f0867ad
---> 592017d223df
Step 4/9 : ENV FLASK_RUN_HOST 0.0.0.0
--->Running in 66fe79634af7
Removing intermediate container 66fe79634af7
---> dbffc8d8f7ac
step 5/9 : RUN apk add--no-cache gcc musl-dev linux-headers
--->Running in 86d3ca6798a4
fetch http://dl-cdn.alpinelinux.org/alpine/v3.12/main/x86_64/APKINDEX.tar.gz
fetch http://dl-cdn.alpinelinux.org/alpine/v3.12/community/x86_64/APKINDEX.tar.gz
(1/13)Installing libgcc(9.3.0-r2)
(2/13) Installing libstdc++(9.3.0-r2)
(3/13) Installing binutils(2.34-r1)
(4/13) Installing gmp (6.2.0-r0)
(5/13) Installing isl(0.18-r0)
(6/13) Installing libgomp (9.3.0-r2)
(7/13) Installing libatomic(9.3.0-r2)
(8/13) Installing libgphobos(9.3.0-r2)
(9/13) Installing mpfr4(4.0.2-r4)
(10/13) Installing mpci (1.1.0-r1)
```

(11/13) Installing gcc (9.3.0-r2)
(12/13) Installing linux-headers(5.4.5-r1)
(13/13) Installing musl-dev (1.1.24-r9)
Executing busybox-1.31.1-r16.trigger
OK: 153 MiB in 48 packages
Removing intermediate container 86d3ca6798a4
---> b2ad3 a89c2f6
step 6/9 : COPY requirements.txt requirements.txt
---> 47e05d81e5b1
Step 7/9 : RUN pip install-r requirements.txt
--->Running in 0539c0d29caa
collecting sk
Downloading sk-0.0.1.tar.gz(1.8 KB)
Collecting redis
Downloading redis-3.5.3-py2.py3-none-any.whl (72KB)
Building wheels for collected packages: sk
Building wheel for sk (setup.py): started
Building wheel for sk(setup.py): finished with status 'done'
Created wheel for sk: filename=sk-0.0.1-py3-none-any,whl size=1883 sha256=8efclf0117b7d24c806628be64fc3b595a86c3a5340fb3e9e720c6a6804d33e5
Stored in directory:/root/.cache/pip/wheels/b9 /35/d5/d1866b0eccb525fdc0b67abbbbb649af280a71917f070cb684
Successfully built sk
Installing collected packages:sk,redis
Successfully installed redis-3.5.3 sk-0.0.1
Removing intermediate container 0539c0d29caa
--->8576f9346b85
step 8/9 : COPY...
---> 2b97f3a0833e
step 9/9 : CMD ["flask","run"]
---> Running in 2bd23bef3d27
Removing intermediate container 2bd23bef3d27
---> 7f69dfe380d2
Successfully built 7f69dfe380d2
Successfully tagged composetest_web: latest
WARNING: Image for service web was built because it did not already exist. To rebuild this image you must use 'docker-compose build' or 'docker-compose up--build'.
Pulling redis (redis: alpine)...
alpine:Pulling from library/redis
df20fa9351 al: Already exists
9b8c029ceab5: Pull complete
e983a1eb737a: Pull complete

完成后可以通过浏览器确认结果，打开浏览器输入"http://ip:5000"，可以看到一条消息（图4-18）。

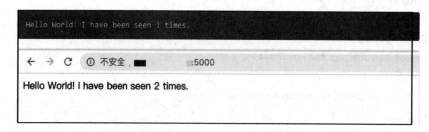

图4-18 通过浏览器访问flask

三、Swarm 集群管理

Docker Swarm 是 Docker 的集群管理工具。它将 Docker 主机池转变为单个虚拟 Docker 主机。Docker Swarm 提供了标准的 Docker API，所有任何已经与 Docker 守护程序通信的工具都可以使用 Swarm 轻松地扩展到多个主机。

支持的工具包括但不限于以下各项：Dokku、Docker Compose、Docker Machine 和 Jenkins。

如图4-19所示，swarm集群由管理节点（manager）和工作节点（work node）构成。

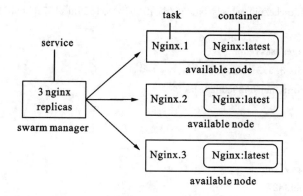

图4-19 Docker Swarm 的工作原理

swarm mananger：负责整个集群的管理工作，包括集群配置、服务管理等所有跟集群有关的工作。

work node：即图中的 available node，主要负责运行相应的服务来执行任务（task）。

以下示例，均以 Docker Machine 和 virtualbox 进行介绍，确保主机已安装 virtualbox。

1. 创建 swarm 集群管理节点（manager）

创建 docker 机器，如下：

```
[root@localhost heathcliff]# docker-machine create -d virtualbox swarm-manager
Running pre-create checks...
Creating machine
(swarm-manager) Copying /root/.docker/machine/cache/boot2docker.iso to /root/.docker/machine/machines/swarm-manager/boot2docker.iso...
(swarm-manager) Creating VirtualBox VM...
(swarm-manager) Creating SSH key...
(swarm-manager) Starting the VM...
(swarm-manager) Check network to re-create if needed...
(swarm-manager) Waiting for an IP...
Waiting for machine to be running, this may take a few minutes...
Detecting operating system of created instance...
Waiting for SSH to be available...
Detecting the provisioner
Provisioning with boot2 docker.
Copying certs to the Local machine directory...
Copying certs to the remote machine.
Setting Docker configuration on the remote daemon...
Checking connection to Docker.
Docker is up and running!
To see how to connect your Docker Client to the Docker Engine running on this virtual machine, run: docker-machine env swarm-manager
[root@localhost heathcliff]# docker-machine ls
NAME            ACTIVE    DRIVER       STATE      URL                         SWARM    DOCKER     ERRORS
swarm-manager   -         virtualbox   Running    tcp://192.168.99.101:2376            v19.03.12
```

初始化 swarm 集群，进行初始化的这台机器，就是集群的管理节点，如下：

```
docker-machine ssh swarm-manager
docker swarm init --advertise-addr 192.168.99.101 #这里的 IP 为创建机器时分配的 ip.
```

如：

```
[root@localhost heathcliff]# docker-machine ssh swarm-manager
('>')
/)  TC (\          Core is distributed with ABSOLUTELY NO WARRANTY
(/-_--_-\)              www.tinycorelinux.net

docker@swarm-manager:~$ docker swarm init --advertise-addr 192.168.99.101
Swarm initialized: current node (ohdn280dq7bbz4u4wfm6xplfy) is now a manager.
To add a worker to this swarm, run the following command:
docker swarm join --token SWMTKN-1-2f6sukmf36scwcakr0mxonepooaqhx6se0jmguwvc0tsuzbu3r-6jam0ry83ixljmm8xl0dsxvny 192.168.99.101:2377
To add a manager to this swarm, run 'docker swarm join -token manager' and follow the instructions.
```

以上输出,证明已经初始化成功。需要把以下这行复制出来,在增加工作节点时会用到:

docker swarm join --token SWMTKN-1-2f6sukmf36scwcakr0mxonepooaqhx6se0jmguwvc0tsuzbu3r-6jam0ry83ixljmm8xl0dsxvny 192.168.99.101:2377

2. 创建 swarm 集群工作节点(worker)

这里直接创建好两台机器:swarm-worker1 和 swarm-worker2。

[root@localhost heathcliff]#docker-machine create -d virtualbox swarm-worker1
Running pre-create checks...
Creating machine...
(swarm-worker1) Copying /root/.docker/machine/cache/boot2docker.iso to /root/.docker/machine/machines/swarm-worker1/boot2docker.iso...
(swarm-worker1) Creating virtualBox VM...
(swarm-worker1) Creating SSH key...
(swarm-worker1) Starting the VM...
(swarm-worker1) Check network to re-create if needed...
(swarm-worker1) Waiting for an IP...
Waiting for machine to be running,this may take a few minutes...
Detecting operating system of created instance...
Waiting for SSH to be available...
Detecting the provisioner...
Provisioning with boot2docker...
Copying certs to the local machine directory...
Copying certs to the remote machine...
Setting Docker configuration on the remote daemon...
Checking connection to Docker...
Docker is up and running!
To see how to connect your Docker Client to the Docker Engine running on this virtual machine,run:docker-machine env swarm-worker1

[root@localhost heathcliff]#docker-machine create -d virtualbox swarm-worker2
Running pre-create checks...
Creating machine...
(swarm-worker2) Copying /root/.docker/machine/cache/boot2docker.iso to /root/.docker/machine/machines/swarm-worker2/boot2docker.iso...
(swarm-worker2) Creating virtualBox VM...
(swarm-worker2) Creating SSH key...
(swarm-worker2) Starting the VM...
(swarm-worker2) Check network to re-create if needed...
(swarm-worker2) Waiting for an IP...

```
Waiting for machine to be running, this may take a few minutes...
Detecting operating system of created instance...
Waiting for SSH to be available...
Detecting the provisioner...
Provisioning with boot2docker...
Copying certs to the local machine directory...
Copying certs to the remote machine...
Setting Docker configuration on the remote daemon...
Checking connection to Docker...
Docker is up and running!
To see how to connect your Docker Client to the Docker Engine running on this virtual machine, run: docker-machine env swarm-worker2
```

```
[root@localhost heathcliff]# docker-machine ls
NAME            ACTIVE   DRIVER       STATE     URL                         SWARM   DOCKER     ERRORS
swarm-manager   -        virtualbox   Running   tcp://192.168.99.101:2376           v19.03.12
swarm-worker1   -        virtualbox   Running   tcp://192.168.99.103:2376           v19.03.12
swarm-worker2   -        virtualbox   Running   tcp://192.168.99.102:2376           v19.03.12
```

分别进入两台机器里,指定添加至上一步中创建的集群,这里会用到上一步复制的内容,如下:

```
[root@localhost heathcliff]# docker-machine ssh swarm-worker2
('>')
/)TC (\              Core is distributed with ABSOLUTELY NO WARRANTY.
(/-_- -_-\)                  www.tinycorelinux.net
docker@swarm-worker2:~$ docker swarm join --token SWMTKN-1-2f6sukmf36scwcakr0mxonepooaqh
x6se0jmguwvc0tsuzbu3r-6jam0ry83ixljmm8xl0dsxvny 192.168.99.101:2377
This node joined a swarm as a worker.
```

```
[root@localhost heathcliff]# docker-machine ssh swarm-worker1
('>')
/)TC (\              Core is distributed with ABSOLUTELY NO WARRANTY.
(/-_- -_-\)                  www.tinycorelinux.net
docker@swarm-worker1:~$ docker swarm join --token SWMTKN-1-2f6sukmf36scwcakr0mxonepoo
aqhx6se0jmguwvc0tsuzbu3r-6jam0ry83ixljmm8xl0dsxvny 192.168.99.101:2377
This node joined a swarm as a worker.
```

以上数据输出说明已经添加成功。

上图中,由于上一步复制的内容比较长,会被自动截断,实际上在图中运行的命令如下:

```
docker swarm join --token SWMTKN-1-2f6sukmf36scwcakr0mxonepooaqhx6se0jmguwvc0tsuzbu3r-6jam0ry83ixljmm8xl0dsxvny 192.168.99.101:2377
```

3. 查看集群信息

进入管理节点，执行 docker info 可以查看当前集群的信息：

```
[root@localhost heathcliff]# docker-machine ssh swarm-manager
       ('>')
      /)TC(\         Core is distributed with ABSOLUTELY NO WARRANTY.
     (/-_--_-\)             www.tinycorelinux.net
docker@swarm-manager:~$ docker info
Client:
 Debug Mode: false
Server:
 Containers: 0
  Running: 0
  Paused: 0
  Stopped: 0
 Images: 0
 Server Version: 19.03.12
 Storage Driver: overlay2
  Backing Filesystem: extfs
  Supports d type: true
  Native Overlay Diff: true
 Logging Driver: json-file
 Cgroup Driver: cgroupfs
 Plugins:
  Volume: local
  Network: bridge host ipvlan macvlan null overlay
  Log: awslogs fluentd gcplogs gelf journald json-file local logentries splunk syslog
 Swarm: active
  NodeID: ohdn280dq7bbz4u4wfm6xplfy
  Is Manager: true
  ClusterID: ebelai2kwa8Lh8960kx7omc26
  Managers: 1
  Nodes: 3
  Default Address Pool: 10.0.0.0/8
  Subnetsize: 24
  Data Path Port: 4789
  Orchest ration:
   Task History Retention Limit: 5
```

```
Raft:
    Snapshot Interval: 10000
    Number of old Snapshots to Retain: 0
    Heartbeat Tick: 1
```

可以看出当前运行的集群中有 3 个节点,其中有 1 个是管理节点(Managers:1;Nodes:3)。

4. 部署服务到集群中

注意:跟集群管理有关的任何操作,都是在管理节点上操作的。

以下例子是在一个工作节点上创建一个名为 helloworld 的服务,这里是随机指派给一个工作节点:

```
docker@swarm-manager:~ $ docker service create --replicas 1 --name helloworld alpine ping docker.com
j8b0qnadgnkwy5mh85cf9cs8u
overall progress: 1 out of 1 tasks
1/1: running   [==================================================>]
verify: Service converged
```

5. 查看服务部署情况

查看 helloworld 服务运行在哪个节点上,可以看到目前是在 swarm-worker1 节点,如下:

```
docker@swarm-manager:~ $ docker service ps helloworld
ID           NAME          IMAGE           NODE            DESIRED STATE   CURRENT STATE            ERROR   PORTS
gynz3vr2szgb  helloworld.1  alpine:latest   swarm-manager   Running         Running about a minute ago
```

查看 helloworld 部署的具体信息:

```
docker@swarm-manager:~ $ docker service inspect --pretty helloworld
ID:                j8b0qnadgnkwy5mh85cf9cs8u
Name:              helloworld
Service Mode:      Replicated
Replicas:          1
Placement:
UpdateConfig:
 Parallelism:      1
 on failure:       pause
 Monitoring Period: 5s
 Max failure ratio: 0
```

```
Update order:      stop-first
Rollbackconfig:
Parallelism:    1
on failure:     pause
Monitoring Period:5s
Max failure ratio:0
Rollback order:    stop-first
ContainerSpec:
Image:      alpine:latest@sha256:185518070891758909c9f839cf4ca393ee977ac378609f700f60a771a2dfe321
Args:       ping docker.com
Init:       false
Resources:
Endpoint Mode:vip
```

6. 扩展集群服务

将上述的 helloworld 服务扩展到两个节点：

```
docker@swarm-manager:~ $ docker service scale helloworld=2
helloworld scaled to 2
overall progress:2 out of 2 tasks
1/2:running [==================================>]
2/2:running [==================================>]
verify:Service converged
```

可以看到已经从一个节点扩展到两个节点：

```
docker@swarm-manager:~ $ docker service ps helloworld
ID              NAME           IMAGE           NODE             DESIRED STATE   CURRENT STATE         ERROR
gynz3vr2szgb    helloworld.1   alpine:latest   swarm-manager    Running         Running 4 minutes ago
tjkr7xem67hf    helloworld.2   alpine:latest   swarm-worker2    Running         Running 59 seconds ago
```

7. 删除服务

```
docker@swarm-manager:~ $ docker service rm helloworld
helloworld
```

查看是否已删除：

```
docker@swarm-manager:~ $ docker service ps helloworld
no such service:helloworld
```

8. 滚动升级服务

以下实例将介绍 redis 版本如何滚动升级至更高版本。

创建一个 3.0.6 版本的 redis，如下：

```
docker@swarm-manager:~ $ docker service create --replicas 1 --name redis --update-delay 10s redis:3.0.6
n7nbs52o9hxf5gv32s9g3gur9
overall progress: 1 out of 1 tasks
1/1: running [==================================================>]
verify: Service converged
docker@swarm-manager:~ $ docker service ps redis
ID              NAME      IMAGE         NODE            DESIRED STATE   CURRENT STATE            ERROR    PORTS
xuwxiugvh75e    redis.1   redis:3.0.6   swarm-worker2   Running         Running 21 seconds ago
```

滚动升级 redis，如下：

```
docker@swarm-manager:~ $ docker service update --image redis:3.0.7 redis
redis
overall progress: 1 out of 1 tasks
1/1: running [==================================================>]
verify: Service converged
docker@swarm-manager:~ $   docker service ps redis
ID              NAME         IMAGE         NODE            DESIRED STATE   CURRENT STATE              ERROR
yjlq6itdgqib    redis.1      redis:3.0.7   swarm-manager   Running         Running about a minute ago
xuwxiugvh75e    \_redis.1    redis:3.0.6   swarm-worker2   Shutdown        Shutdown about a minute ago
```

可以看出，redis 的版本已从 3.0.6 升级到 3.0.7，说明服务已经升级成功。

9. 停止某个节点接收新的任务

查看所有的节点：

```
docker@swarm-manager:~ $ docker node ls
ID                            HOSTNAME         STATUS   AVAILABILITY   MANAGER STATUS   ENGINE VERSION
ohdn280dq7bbz4u4wfm6xplfy *   swarm-manager    Ready    Active         Leader           19.03.12
ooiq5qbykumqvxsisval9twig     swarm-worker1    Ready    Active                          19.03.12
8d9xcr5lo3a75dgjag60z594l     swarm-worker2    Ready    Active                          19.03.12
```

可以看到目前所有的节点都是 Active，可以接收新的任务分配。

停止节点 swarm-worker1：

```
docker@swarm-manager:~$ docker node update --availability drain swarm-worker1
swarm-worker1
docker@swarm-manager:~$ docker node ls
ID                          HOSTNAME        STATUS  AVAILABILITY  MANAGER STATUS  ENGINE VERSION
ohdn280dq7bbz4u4wfm6xplfy * swarm-manager   Ready   Active        Leader          19.03.12
ooiq5qbykumqvxsisval9twig   swarm-worker1   Ready   Drain                         19.03.12
8d9xcr5lo3a75dgjag60z594l   swarm-worker2   Ready   Active                        19.03.12
```

注意：swarm-worker1 状态变为 Drain，不会影响到集群的服务，只是 swarm-worker1 节点不再接收新的任务，集群的负载能力有所下降。

可以通过以下命令重新激活节点：

```
docker@swarm-manager:~$ docker node update --availability active swarm-worker1
swarm-worker1
docker@swarm-manager:~$ docker node ls
ID                          HOSTNAME        STATUS  AVAILABILITY  MANAGER STATUS  ENGINE VERSION
ohdn280dq7bbz4u4wfm6xplfy * swarm-manager   Ready   Active        Leader          19.03.12
ooiq5qbykumqvxsisval9twig   swarm-worker1   Ready   Active                        19.03.12
8d9xcr5lo3a75dgjag60z594l   swarm-worker2   Ready   Active                        19.03.12
```

主要参考文献

北京红亚华宇科技有限公司,2019.云平台安装配置手册[R].北京:北京红亚华宇科技有限公司.

顾军林,徐义晗,米洪,等,2019.虚拟化与网络存储技术[M].北京:人民邮电出版社.

青岛英谷教育科技股份有限公司,2018.云计算与虚拟化技术[M].西安:西安电子科技大学出版社.

史律,薛飞,许建铭,等,2020.虚拟化技术实现与应用[M].北京:电子工业出版社.

天津滨海迅腾科技集团有限公司,2019.Docker虚拟化技术入门与实战[M].天津:天津大学出版社.

王鹏,黄焱,安俊秀,等,2019.云计算与大数据技术[M].北京:人民邮电出版社.

肖睿,2017.Docker容器与虚拟化技术[M].北京:水利水电出版社.

中智讯(武汉)科技有限公司,2015.虚拟化与云计算开发指南[R].武汉:中智讯(武汉)科技有限公司.

docker hub(docker镜像网站)[EB/OL].(2021-04-15)[2021-04-19].https://hub.docker.com/.

docker中文社区[EB/OL].(2021-04-19)[2021-04-19].https://www.docker.org.cn/.

Kernel Virtual Machine[EB/OL].(2021-03-25)[2021-04-19].https://www.linux-kvm.org.

附录　CentOS 中 NFS(网络文件系统)的配置方法

(1)创建 NFS 的共享目录(如"/export/nfs_sr"),并开放读、写、执行权限。

```
[root@centos7 ~] cd /
[root@centos7 ~] mkdir export
[root@centos7 ~] chmod 777 export
[root@centos7 ~] cd export
[root@centos7 ~] mkdir nfs_sr
[root@centos7 ~] chmod 777 nfs_sr
```

(2)安装 NFS 相关软件。

```
[root@centos7 ~] yum install nfs-utils
[root@centos7 ~] yum install rpcbind
```

(3)将 NFS 服务配置为开机自启动。

```
[root@centos7 ~] systemctl enable nfs
[root@centos7 ~] systemctl enable rpcbind
```

(4)启动 NFS 服务。

```
[root@centos7 ~] systemctl start rpcbind
[root@centos7 ~] systemctl start nfs-server
```

(5)修改 NFS 配置文件,将共享目录添加到 NFS 的 export(输出)列表中去,＊表示所有 IP 地址都可访问该目录。

```
[root@centos7 ~] gedit /etc/exports
/export/nfs_sr/ *(rw,async,no_root_squash,no_subtree_check)
```

常用参数说明如下。

rw:表示可读写。

sync:表示同步写。

no_root_squash:NFS 客户端连接服务端时如果使用的是 root,那么对服务端分享的目录来说,也拥有 root 权限。

no_subtree_check:共享/usr/bin 之类的子目录时,NFS 不检查父目录的权限。

(6)刷新,让 exports 配置信息立即生效,并验证配置内容是否生效。

```
[root@centos7 ~] exportfs -a
[root@centos7 ~] exportfs -rv
exporting *:/export/nfs_sr
```

(7) 如果新添加了 NFS 共享目录，需要重复第一步操作，新建共享目录，并开放其读、写、执行权限，然后重启 NFS 服务。

```
[root@centos7 ~] systemctl restart nfs-server
[root@centos7 ~] systemctl restart rpcbind
```

(8) 关闭防火墙。

查看防火墙状态命令如下。

```
[root@centos7 ~]# systemctl status firewalld
```

如果看到防火墙状态是 active(running)，就意味着防火墙是打开状态。

为了保证实验的顺利进行，建议临时关闭 NFS 服务器上的防火墙，或永久关闭防火墙。

临时关闭防火墙服务的命令如下。

```
[root@centos7 ~]# systemctl stop firewalld.service
```

永久关闭防火墙服务的命令如下。

```
[root@centos7 ~]# systemctl disable firewalld.service
```

(9) 查看 NFS 服务器输出的目录列表(这里 192.168.1.15 是 NFS 服务器的 IP 地址)。

```
[root@centos7 ~]# showmount -e 192.168.1.66
Export list for 192.168.1.66:
/export/nfs_sr *
```